Innovation Strategies in New Product Development

European University Studies
Europäische Hochschulschriften
Publications Universitaires Européennes

Series V
Economics and Management

Reihe V   Série V
Volks- und Betriebswirtschaft
Sciences économiques, gestion d'entreprise

Vol./Bd. 3390

PETER LANG
Frankfurt am Main · Berlin · Bern · Bruxelles · New York · Oxford · Wien

Tanja Rajkovič

Innovation Strategies in New Product Development

Balancing Technological, Marketing and Complementary Competencies of a Firm

PETER LANG
Internationaler Verlag der Wissenschaften

Bibliographic Information published by the Deutsche Nationalbibliothek
The Deutsche Nationalbibliothek lists this publication in the Deutsche Nationalbibliografie; detailed bibliographic data is available in the internet at http://dnb.d-nb.de.

D 30
ISSN 0531-7339
ISBN 978-3-631-61963-6

© Peter Lang GmbH
Internationaler Verlag der Wissenschaften
Frankfurt am Main 2011
All rights reserved.

All parts of this publication are protected by copyright. Any utilisation outside the strict limits of the copyright law, without the permission of the publisher, is forbidden and liable to prosecution. This applies in particular to reproductions, translations, microfilming, and storage and processing in electronic retrieval systems.

www.peterlang.de

Acknowledgments

Along this journey I was fortunate to have had wonderful support of many people. I would first like to thank my mentor prof. dr. Janez Prašnikar from the Faculty of Economics, University of Ljubljana (Slovenia), and co-mentor prof. dr. Lubomir Lizal from University of Prague (Czech Republic) for their guidance and for sharing their knowledge with me.

I am thankful for the invaluable advice also to doc. dr. Polona Domadenik, prof. dr. Vesna Žabkar from Faculty of Economics, University of Ljubljana, prof. dr. Maja Vehovec from The Institute of Economics, Zagreb (Croatia), and to prof. dr. Christian Ringle from the Institute for Industrial Management, University of Hamburg (Germany) for helping me out over the e-mails with SmartPLS related dilemmas. To all my other colleagues at the Faculty of Economics, University of Ljubljana, I would like to express my thanks for all the hints and words of wisdom.

To my family and friends, especially my dad Vladislav and brother Uroš, I am immensely grateful for their incessant support and for always believing in me.

This research was supported by a grant from the CERGE-EI Foundation under a program of the Global Development Network. All opinions expressed are those of the author and have not been endorsed by CERGE-EI or the GDN.

Contents

1	**Introduction**		11
	1.1	Subject and objective	14
	1.2	Methodology	15
	1.3	Structure	16
2	**Innovation and theory of endogenous growth**		19
3	**Innovation and firm competitiveness**		25
	3.1	Innovation in high technology versus low- and medium-technology industries	27
	3.2	Specifics of service innovation	29
	3.3	Innovative performance	30
		3.3.1 Incremental and radical innovation	32
		3.3.2 Technological and market turbulence	34
	3.4	Lisbon strategy and innovative activity in the European Union	35
		3.4.1 Community Innovation Survey – Slovenia	38
		3.4.2 Technology leaders and followers	39
4	**Theories of competitive advantage**		43
	4.1	Resource-based theory	43
	4.2	Dynamic capabilities theory	45
	4.3	Competence-based theory	46
5	**Firm competencies and competitive advantage**		49
	5.1	Competencies as a source of competitive advantage	53

5.2	Competencies as drivers of innovation in the R&D function	57
5.3	Previous empirical studies	60
5.4	Technological competencies	65
5.5	Marketing competencies	67
5.6	Complementary competencies	69

6 Model of competencies as antecedents of innovative performance and subsequent effect on business performance 73

6.1	Operational model	73
6.2	Methodology and survey design	76
6.3	Variables	81
6.4	Data	85
	6.4.1 General company data	88
	6.4.2 R&D activities and the production function	90
	6.4.3 Aggregate R&D company data	92
6.5	Innovative performance based clustering	94
6.6	Structural models	101
6.7	Structural models of competencies and innovative performance	103
6.8	Extensions of the baseline model	113
	6.8.1 Complementary competencies as interaction between technological and marketing competencies	113
	6.8.2 Extension of the baseline model for business performance	117
	6.8.3 Sampling bias	120

	6.8.4 Moderating effects of environmental turbulence	128
7	**Conclusion**	137
	7.1 Contribution to theory and practice	139

References 143

Appendices

    A Comparison of the contemporary strategic
management approaches 165

    B Studies aimed at developing the theory of competencies 166

    C Questionnaire 170

    D Dendrogram 181

    E Comparison of object classification with hierarchical
Ward's procedure and K-means method into 3 clusters 182

    F PLS structural model analysis for the incremental innovation
model of innovative performance 183

    G PLS structural model analysis for the radical innovation
model of innovative performance 185

    H PLS structural model analysis for the trend setting/market
leadership model of innovative performance 187

1 Introduction

In a dynamic environment companies constantly strive for ways to differentiate themselves from their competitors and in so doing aim to benefit from the thus-created competitive advantage. Innovation activity is recognized for creating such opportunities especially as companies are unanimously reaching for universally high standards of products and services, entering alliances, participating in industry consolidation and building broadly matching global brands, as well as distribution capabilities.

For the last three years the economic focus persists to be on the financial crises that have swept over the global economy. The awareness of the crises most notably arose with the 2008 collapse of firms in the investment services industry, particularly with the bankruptcy of the Lehman Brothers Holdings Inc., the 4th largest investment bank in the USA at that time. Speculations on the mortgage markets, increase in risky loans and weak capital structure of banks led to difficulties with their liquidity (for more on the topic see Diamond and Rajan (2009)). It is very much clear now that the current financial crises are in their origins also banking crises (Gertler, 2010).

What this brought about, is the so called "great recession" with severe consequences also for the non-financial sectors that are enduring a significant slowdown. Financial crises, also credit crises, inevitably cause deep and prolonged asset market collapses, significant falls in employment and output and the explosion of the real value of the government debt (Reinhart & Rogoff, 2010). We are now witnessing all of these consequences while the recovery seems uncertain. Furthermore, it is becoming obvious that during the economic expansion many companies chose to take part in the speculations on the financial markets as a part of their financing strategy, distancing themselves from their core businesses. Now they are not only affected by the rising prices of inputs and a slump in demand but have also witnessed their financial assets, on which they have become more and more reliant, disappear. There is also less capital available for investments.

With daily news reports on new firm bankruptcies and rising levels of unemployment, one may easily get the impression that the poor performance of the real sector is to blame. Interestingly enough, this is not the case as the rules in the real sector remain unchanged. In the wake of the crisis, firms were indeed first forced to pay close attention to managing financial risks, but as this has slowly become accepted as an inevitable reality, new growth and strategic guidelines are being sought after. Even though the initial reaction of many firms resulted in shrinking production capacity, downsizing labor force, reducing discretionary spending and conserving cash, this can only be a short term strategy (Reeves & Deimler, 2010).

In terms of long term strategy, the recent McKinsey Global Survey on Innovation and Commercialization (McKinsey&Company, 2010) still confirms the central role of innovation. It sets forth that innovation has again become a priority for firms as they begin to refocus on growth. 84% of the 2,240 surveyed executives from around the world claim that innovation is extremely or very important to their companies' growth strategy. Two years into the recession, the percentage of respondents who find their companies are still better at innovation that their peers, remains constant at 55% (compared to survey results from 2008). Innovation has not lost its strategic role.

In terms of academic research prior to the current recession, the link between innovation and growth on a national level has been extensively researched in the literature from both theoretical and empirical perspectives. The theory of endogenous growth explores technological progress by focusing on human capital, innovation motives in the form of patents that enable monopolistic gains, and the significance of the spillover effect (Barro & Sala-i-Martin, 1997; Lucas, 1988; Romer, 1986). Authors Tong and Xu (2006) demonstrate with an extended model of endogenous growth adjusted for transition economies that in order for these countries to catch up with the developed economies it is important to undertake and undergo both technological and institutional changes. The central role of R&D investments confirms a large number of macroeconomic studies that are based on the models of endogenous growth (Griffith et al., 2004; Bassanini & Scarpetta, 2001; Guellec & van Pottelsberghe de la Potterie, 2001). It is on these theoretical grounds that the Lisbon strategy for promoting economic development in the European Union is also based (Kok, 2004).

Newer studies on economic growth analyze total factor productivity (TFP). As two of the biggest constituents of TFP are considered to be technology growth and efficiency, this can explain 60% of the difference in national income (Klenow & Rodriguez-Clare, 1997) Similar results were obtained by Easterly and Levine (2001) who showed that TFP accounts for around half of the growth of the real GDP per capita and 90% of variations among countries.

Grossman and Helpman (1994) further explain two approaches to growth, the first being growth by imitating – which is typical for the developing countries – and the second being growth by innovating, characteristic of the developed economies. Barro and Sala-i-Martin (1997) state that capital accumulation and technology transfer can be a successful approach to accelerated growth in less developed economies since the process of imitating requires less substantial investments. On the other hand, developed countries can maintain their advantage only by means of perpetual innovating processes. A large body of research carried out on the level of firms confirms the positive effect of innovation on productivity (Wakelin,

2001; Mairesse & Sassenou, 1991 Griliches & Mairesse, 1983) and the market value of firms (Nagaoka, 2006; Bosworth & Rogers, 2002; Blundell et al., 1999; Hall, 1999; Bosworth & Mahdian, 1999).

Technological innovation may appear to be the only means by which firms compete, however, national action plans for driving innovation list a variety of policies. The EU's report entitled 'Creating an Innovative Europe' proposes a strategy focusing on the creation of innovation-friendly markets, strengthening R&D resources, and increasing structural mobility as well as fostering a culture which celebrates innovation. In order to create an innovation-friendly market in which to launch new products and services, certain actions are recommended, regarding harmonized regulation, an ambitious use of standards, the driving of demand through public procurement and a competitive intellectual property rights regime (Aho et al., 2006).

In general, innovative companies should be more successful than their non-innovative counterparts (Griffith et al., 2004; Tether, 2002). According to this line of thinking, the main reason for lower long-term growth in Europe, compared with the USA, is considered to be the lower R&D expenditure of governments and companies (Gassmann & von Zedtwitz, 1999; Sapir, 2003; von Zedtwitz, 2004). The gap between the US and Europe in this field has even increased in recent decades (Sapir, 2003).

The analysis of data from polls on innovation and R&D activities in 2992 Slovenian firms from manufacturing and service sectors in the year 2002 finds that innovative companies constitute only 21% of the total number. There is a positive bias for large companies, companies that are partially owned by foreigners, and for export-oriented companies (Stanovnik & Kos, 2005). Innovation and R&D expenditures have been stagnating for several years now and are lower than in developed European countries. The majority of Slovenian manufacturers (66%) employ medium-low or low technology according to OECD classification. The comparative gap with some European countries (Austria, Finland) is particularly large in classes of companies that use medium-high and medium-low technology. The share of external expenditure accounted for by R&D in innovation expenditure is less than 10%. There is weak cooperation with other companies in the formation of technological knowledge formation and in drawing knowledge from the academic environment (Prašnikar, 2006).

The strategic management literature and theories of competitive advantage present a more extensive perspective on means of competition and do not merely focus on technological innovation. When companies compete in a dynamic environment, the product-centred perspective on strategy might explain a firm's current competitive advantage. However, this perspective does not facilitate a strategy making process that creates competitive advantage in the future (Fowler et al., 2000). The source of a firm's competitive advantage rather rests on its capabilities and competencies (Song et al., 2005; Lynskey, 1999; Prahalad & Hamel, 1990). Since the 1980s, three approaches to competitive strategy that firms should pursue have emerged, namely: the resource-based view (Barney, 1991; Wernerfelt, 1984), the competence-based perspective (Prahalad & Hamel, 1990) and the dynamic capabilities approach (Teece et al., 1997).

Capabilities are defined as continuous patterns of activities that utilize a firm's resources to generate products for the market, and are largely industry specific. They are intangible assets that nonetheless determine the application of other tangible and intangible resources. (Sanchez, 2004; Hafeez et al., 2002). Competencies, on the other hand, refer to the ability to utilize resources that spread across multiple functions, products and markets in a sustainable and synchronized manner. Their main constituents are capabilities and a portfolio of capabilities respectively. Competencies, namely a network of capabilities and other firm resources, differ from company to company, yet represent a broader, more general perspective on strategy and are not industry related. If a company strives to accomplish strategic goals, it needs to develop the competencies dynamically so as to be able to adjust to changes, both in the external environment and within the firm. Sustainability is established through the retention of organizational focus. Accordingly, competencies have strategic potential for seizing opportunities and neutralizing threats posed by competitors (Sanchez, 2004).

Within the context of the current economic downturn especially, innovation, competencies and competitive advantage still comprise a firm's core. Jeffrey Immelt, CEO of GE, which is the world's largest industrial firm, summarized his thoughts on innovation in the current unfavourable economic climate in the following way: "Companies and countries that really play offence vis-à-vis technology and innovation are going to come out ahead" (The Economist, 2008). Therefore, innovation and competence building should constantly remain high among the priorities, yet an understanding of these concepts is needed in order to be able to reap maximum benefits.

1.1 Subject and objective

A number of empirical studies (Hagedoorn & Cloodt, 2007; Song et al., 2005; Wang et al., 2004) have tried to differentiate the various sources of superior firm performance in terms of different elements of core competencies, and thus provide an insight into the underlying determinants of innovation and, consequently, innovative performance. Moreover, a few empirical studies can be found that examine the major constituents of core competencies and their differentiated influences on overall firm performance (Wang et al., 2004). Such research is needed to achieve an in-depth understanding of how and why core competencies contribute to firm performance in contingent contexts; what is more, in order to adapt quickly and effectively to the increasingly changing nature of both internal and external business environments, without focusing solely on the technological aspect of innovation activities.

This is the main focal point of the research presented in this book from the viewpoint of Slovenian manufacturing firms. The underlying research question is also how companies' competitive positions are reflections of their competencies and innovative performance, with the working assumption that Slovenia takes the role of a technology follower.

The purpose of this work is to validate an operational model of innovative performance based on three groups of competencies contributing to new product development – technological, marketing and complementary – and also to examine the relationships with respect to business performance.

1.2 Methodology

The operational model is constructed drawing from a synthesis of the literature in the field of theories of competitive advantage and innovation. Due to the novelty and specifics of the developed model, it cannot be tested using existing datasets. Therefore, one of the goals herein was to devise a survey that can be used for multi-industry studies of firm competencies. The nature of competencies makes it possible to compare companies from different industries (multi-industry analysis) since they are neither industry specific nor bound to particular products and companies (Sanchez, 2004).

To test the hypotheses and operational model a set of different statistical tools is employed, beginning with a descriptive analysis and describing the sample with aggregate data for different firm characteristics. This is followed by identification of different firm segments using the clustering method. Analyzed are the differences among segments with respect to their competencies and innovative performance. In clustering a two step methodology is applied. This technique proposes improving the segmentation initially obtained via hierarchical clustering methods by additionally applying non-hierarchical methods in order to optimize the classification of the observation set.

The second part of the empirical analysis is dedicated to structural models, where the relationships between competencies, innovative performance and business performance are established. Applied is the Partial Least Squares technique for structural equation modelling (SEM) (Chin & Newsted, 1999; Chin, 1998), more specifically the SmartPLS tool. SEM is a collection of statistical techniques that allows us to examine the set of relationships obtaining between one or more independent and dependent variables. The PLS approach to structural modelling poses minimal demands on measurement scales, sample size and residual distributions. The method has the capacity for both theory testing and theory development.

1.3 Structure

The thesis is essentially divided into two main parts, the first presenting theoretical backgrounds and the second pertaining to the empirical analysis.

In the beginning are presented the main mechanisms and findings of the theory of endogenous growth from the perspective of the role innovation plays in driving economic growth. Firm competitiveness and its main concepts are addressed in terms of strategic importance of innovation. Next are presented the differences in how innovation is regarded in high technology industries compared to low- and medium-technology industries. After laying out the specifics of service innovation, innovative performance as a measure of innovation is discussed, while special emphasis is given to incremental and radical innovation. Special attention is also given to the effects of environmental turbulence with regard to innovation activities.

Innovations have a special role in economic policy and are referred to in the chapter on the Lisbon strategy. Here are presented various findings on national innovation activity, with most attention dedicated to Slovenia. Establishing Slovenia as a technology follower country, this concept is further explored to shed light on what differentiates these countries from technology leaders.

Presentation of the three theories of competitive advantage is followed by a detailed discussion on how firm competencies mitigate the creation of competitive advantage. Specifically are addressed competencies to which innovation and successful new product development can be attributed. Relevant previous empirical studies are listed in order to best illustrate the field. Technological, marketing and complementary competencies are all addressed separately and in more detail, both with their definitions and prevailing measures.

The empirical part starts with an elaboration of the operational model and research question, consisting of further development into 13 hypotheses. The chapter on methodology is complemented by a subchapter on survey design. Subsequently, the variables and data set that were used are presented. This is followed by a descriptive analysis of data and firm segmentation based on innovative performance.

The chapter on the structural model opens with an overview of the method in use. Firstly, four models of innovative performance are tested. The baseline model of innovative performance is modified in further chapters allowing for interaction between technological and marketing competencies, and extended for business performance and tested for sampling bias. Finally, environmental effects are introduced to the model.

The thesis closes with a conclusion discussing the main findings. Addressed are also the main contributions to theory and practice, including the implication in terms of economic downturn of recent years.

2 Innovation and theory of endogenous growth

In his comprehensive and rather unconventional analysis of economic development dating to the first half of the 20th century, Schumpeter refers to innovations as new combinations that are economically more viable than the old way of doing things (Schumpeter, 1983, p. 66). Discontinuous emergence of the new combinations is what in turn drives economic development.

Technological progress can, among other factors affecting long-term growth rates of countries, thus help explain why countries differ dramatically in standards of living. Barro and Sala-i-Martin (1995, p. 4) argue that even small differences in long-term growth rates, when accumulated over a longer period, have much greater consequences for standards of living than the short-term business fluctuations which typically occupy the majority of the attention of macroeconomists.

A new body of research on economic growth arose in the mid-1980's as it was observed that determinants of long-run economic growth are of key importance, surpassing the theories prevailing at that time regarding the mechanics of business cycles or the countercyclical effects of monetary and fiscal policies. Recognition of this represented a starting point for breaking the boundaries of the neoclassical growth model that is characterized by long-term per capita rate being contingent on the rate of exogenous technological progress (Barro & Salla-i-Martin, 1995, p. 12-13). Thus, establishing the determinants of long-term growth within the model brought about endogenous growth models.

Initial research in this field was performed by Romer (1986), Lucas (1988) and Rebelo (1991). In these models it was possible for growth to continue indefinitely since the returns to investment in capital goods, including human capital, do not necessarily diminish with the development of the economies in question.

Researchers have thus included technological development in the models. Technological development is a result of deliberate R&D activity and makes gains of ex-post monopoly power possible. It is the prospect of monopoly profits that motivates R&D investments. If inventive activity and technological advances are continuous, then the long-term growth rate can be positive. The creation of new goods and methods of production can be facilitated by governmental actions, among them taxation, changes in the legal system and the protection of intellectual property rights, infrastructure services, as well as regulations regarding international trade, financial markets and the like. The role of government should therefore not be overlooked.

Newer research further incorporated the diffusion of technology. Unlike the above mentioned technological discovery that is taking place in the leading-edge economies, diffusion of technology makes it possible for follower economies to take part in these advances by means of the cheaper strategy of imitation. Consequently, the diffusion models predict convergence similar to the predictions of the neoclassical growth model.

The manner in which the two models differ technically is presented in what follows. The Solow growth model as a neoclassical growth model demonstrates how saving, population growth and technological progress affect the level of an economy's output and its growth over time. It is built on a basic production function:

$$Y = F(K,L) \tag{1}$$

Equation 1 states that output Y depends on the capital stock K and the labour force L. It is an assumption of the model that the production function witnesses constant returns to scale, which enables the analysis of the quantities entering the model relative to the size of the labour force:

$$\frac{Y}{L} = F\left(\frac{K}{L}, 1\right) \text{ or expressed as } y = f(k) \tag{2}$$

In the Solow model consumption and investment are the two sources of demand for goods. Output per worker y is thus divided between consumption per worker c and investment per worker i:

$$y = c + i \tag{3}$$

The present model is simplified in such a way that it omits government purchases and net exports, assuming a closed economy. A further assumption of the model is that each year people save a fraction s of their income and consume a fraction (1-s). The consumption function can be expressed as follows:

$$c = (1-s) \cdot y \tag{4}$$

where saving rate s assumes values between 0 and 1 and is given. By joining the above equations we obtain:

$$y = (1-s) \cdot y - i \tag{5}$$

$$i = s \cdot y \tag{6}$$

$$i = s \cdot f(k) \tag{7}$$

Investment thus equals saving. Expanding the model for depreciation, that is the constant fraction δ of the stock capital that wears out every year, can be performed by including the equation expressing the impact of investment i and depreciation δk on the annual change in capital stock Δk (Mankiew, 2003, p. 180-185; Barro & Sala-i-Martin, 1997; Solow, 1956):

$$\Delta k = i - \delta \cdot k \tag{8}$$

$$\Delta k = s \cdot f(k) - \delta \cdot k \tag{9}$$

The level of capital stock in the long-run equilibrium of the economy at which investment and depreciation balance is the steady-state level of capital. Growth in the number of workers decreases capital per worker by what is accounted for in the model by the population growth rate n (Mankiew, 2003, p. 201):

$$\Delta k = s \cdot f(k) - (\delta + n) \cdot k \tag{10}$$

Finally, technological progress enters the model in virtue of its effect on the efficiency of labour, which in turn reflects a given society's knowledge about production methods. Labour-augmenting technological progress is denoted by rate g (Mankiew, 2003, p. 208-209):

$$\Delta k = s \cdot f(k) - (\delta + n + g) \cdot k \tag{11}$$

The Solow model posits that once the steady state is reached, the rate of growth of output per worker depends only on the rate of technological progress. It is only technological progress that can explain persistently rising living standards.

Endogenous models, on the other hand, show that incorporating technological progress as a source of growth in the model means improving the production function. An improved production function will result in increased output for the same input value. A production function that includes the effect of technological change can be written as:

$$Y = A \cdot F(K, L) \tag{12}$$

In the equation above, A represents the measure of the current level of technology or the so-called total factor productivity (TFP). Any increase in output Y is not only a consequence of an increase in other production factors – capital K and labour L – but also due to increases in TFP.

Changes in technology are accounted for by the following equation of economic growth:

$$\frac{\Delta Y}{Y} = \alpha \frac{\Delta K}{K} + (1-\alpha)\frac{\Delta L}{L} + \frac{\Delta A}{A} \qquad (13)$$

As previously mentioned, there are three sources of growth in output $\frac{\Delta Y}{Y}$, namely: contribution of capital $\alpha \frac{\Delta K}{K}$, α representing capital's share; contribution of labour $(1-\alpha)\frac{\Delta L}{L}$, and growth in TFP $\frac{\Delta A}{A}$. Unlike growth in output, capital and labour, TFP cannot, as such, be measured directly:

$$\frac{\Delta A}{A} = \frac{\Delta Y}{Y} - \alpha \frac{\Delta K}{K} - (1-\alpha)\frac{\Delta L}{L} \qquad (14)$$

Thus, it represents the change in output that can not be ascribed to changes in inputs. It is computed as a residual and, following Robert Solow (1957), referred to as the Solow residual. Changes in TFP are most often due to increases in knowledge concerning production methods. This explains why any change in TFP or the Solow residual is used as a measure of technological progress. However, TFP can account for any source that changes the relation between the measured inputs and the measured output. Such sources may include a higher quality of education that consequently increases workers' productivity. This also means that the government can affect TFP and growth of output by taking measures that affect factors related to productivity. In the case of education this could be anything from its regulation to changes in state funding. An analysis of sources of growth between 1950 and 1999 in the USA, reveals that increases in capital, labour and TFP have contributed almost equally to economic growth of 3.6% per year, said contributions being 1.2%, 1.3% and 1.1% respectively (Mankiew, 2003, p. 233).

A similar analysis of the economic growth in the "Tigers" of East Asia – Hong Kong, Singapore, South Korea and Taiwan – during the period from 1966 to 1990 revealed that exceptional average annual growth of roughly 7% is somewhat different. The ability of these economies to imitate foreign technologies and improve their own production functions within a short period of time was recognized as a key source of their rapid growth. Therefore, after accounting in the model of economic growth for increases in labour, capital and human capital, only a small portion of the growth was left unexplained. The average growth in TFP was small and similar to that of the USA and, as such, not central to the growth of the "Tigers" in the second half of the 20th century (Young, 1995).

However, it is important to note that innovations do not automatically translate into increased TFP and growth. One may only infer that the more widespread diffusion and adoption of innovation is, the greater the impact on growth and efficiency will be, leading to greater incentives for further innovative activity (Robertson et al., 2009).

3 Innovation and firm competitiveness

Drucker (2007, p. 27-32) defines innovation as a firm's core process and suggests that the best, and possibly the only, way a business can prosper in an environment of rapid change is to innovate and, in so doing, convert change into opportunities. Although distinctive features of a product or service can constitute competitive differentiation advantages for a firm in the marketplace, they can also erode, either due to competitors' actions or changes in customers' preferences. Therefore, firms need to continuously look for new ways to achieve these differentiation advantages (Varadarajan, 2009).

Not all innovations result in success; however, those that do can be a crucial source of competitive advantage. As proposed by Porter (1998b, p. 37-44) firms employ differentiation strategies in order to achieve a competitive advantage by creating a product or service that is perceived as unique. A firm's ability to satisfy customers' needs in this way also implies that a firm can charge a price premium for its products above the industry norm. Product differentiation can be achieved in several ways, including product innovation, technical superiority, product quality and reliability, comprehensive customer service, and unique competitive capabilities (Thompson et al., 2005, p. 149).

Innovation can refer to anything new or novel in either how the company operates or the products it produces. Francis and Bessant (2005, p. 180) classify the four basic types of innovation, of which the first two are prevalent:

- change in terms of changes in what a firm offers via its products/services: product/service innovation,
- innovation in the ways a firm creates and delivers those offerings: process innovation,
- change in the context in which a product/service is applied: market position innovation, and
- change in the underlying industry or business models surrounding the product/service: paradigm innovation.

Luchs (1990) draws the conclusion - based on research in the management field - that those firms able to use innovation to differentiate their products and services from competition in such a way that they are perceived as being of high relative quality, are, on average, twice as profitable as their counterparts when measured in terms of return on investment. However, some innovation initiatives can be dysfunctional and also lead to catastrophic losses.

In a study of Japanese firms, Deshpandé and Farley (2004) demonstrated how corporate culture, customer-orientation and innovativeness are linked to organizational performance measured as relative profitability, size, market share and growth rate. Their findings are based on research spanning a decade and including 12 countries. Innovativeness and customer-oriented marketing appear to have the most positive link with firm performance in all national settings, be it an industrial country or a transition economy. Baldwin and Johnson (1996, p. 800-802) showed, via a sample of 820 Canadian firms, that more innovative firms also have more favourable performance measures, including market share gain and return on investment. This same study also revealed that more innovative firms place a greater emphasis on strategies in key areas – such as management, human resources, marketing, finance, government programs and services, and production efficiencies – than do less innovative firms.

While there is a common consensus regarding the importance of innovation, there is much more disparity in defining which activities actually constitute the innovation process. One of the open questions remains whether R&D is either a necessary or a sufficient condition for innovation. A clear notion is crucial for the understanding of innovation and its impact on a firm's success. Many studies use innovativeness as a synonym for R&D activity. However, a firm can innovate even without engaging in R&D (Gottardi, 1996). Napolitano (1991) studied innovative activity in a much broader sense, not limiting his research to R&D activity alone. In an analysis of 8220 innovative Italian firms, R&D scored only 2.1 out of a possible 6 in terms of its importance as a source of innovation. Other sources of innovation that were rated higher included: purchase of equipment (4.0), design (3.1), employee proposals (2.3), customer requests (2.3), and staff training (2.2). Differences were observed also between industries. Indeed, the importance of these various factors varies considerably by industry. Firms in advanced, science-based industries depend mostly on internal sources of innovation, among them R&D, design, and employee proposals. Mass-production industries are more reliant on staff training and both upstream and downstream activities. Furthermore, firms in the traditional industries (food, textiles, paper, and metal) draw primarily from the purchase of equipment.

A study by Archibugi et al. (1991) similarly suggests that R&D is not very often the source of scientific and technological knowledge (S&T knowledge) that in turn generates innovative activity. For producers of traditional consumer and intermediate goods these sources are design and tooling-up and the purchase of capital goods. Specialized suppliers of intermediate goods report these sources to be equipment, design and tooling-up, R&D, and the acquisition of S&T information sources (professional organizations, technical centres, customers, trade fairs, and exhibitions) and patents and know-how. For mass-production assemblers, design and tooling-up, R&D and acquisition of patents and know-how are the key sources of S&T knowledge. In R&D-based firms, R&D and acquisition of patents and know-how make the most significant contribution to the accumulation of S&T knowledge.

In order to understand innovation, it is equally important to recognize that the scientific and technological activities of a firm require a supporting structure. A firm must finance its activities, deploy physical and human resources, market its products and services, and successfully coordinate all of these activities. It is as important to be aware of the supporting role of these policies as it is to investigate the breadth of the scientific and technological activities in a firm (Baldwin & Johnson, 1996).

As the complexity of technologies and new business practices is increasing simultaneously with the ongoing globalization of markets, many firms are forced to rely on R&D as a source of strategy for long-term growth and sustainability (Mikkola, 2001, p. 433). R&D has two primary roles in achieving superior innovation. Firstly, through new product and process development. Secondly, through effectiveness of R&D management that depends on ability of the R&D department to cooperate smoothly with both marketing and manufacturing departments (Prajogo et al., 2008, p. 620).

The results of a study based on a sample of 74 biotechnology companies in Canada showed that R&D intensity, which was self-reported by the sample firms as the percentage of total revenues going allotted to R&D activity, correlates with patent measures, whereas innovation measures in terms of new product introductions is linked to business performance (Hall & Bagchi-Sen, 2002, p. 238).

3.1 Innovation in high technology versus low- and medium-technology industries

The widely adopted classification of manufacturing industries, which divides them into high-, medium- and low-tech, as also used by the OECD, has recently received much criticism. The classification is based on R&D intensity measured as R&D expenditure, the threshold being at 5% of revenues. Researchers namely oppose equating high R&D intensity with high innovativeness. R&D is, in reality, only one possible way of attaining innovativeness. Additionally, the sectoral approach does not adequately take into account differences at the firm level (Kirner et al. 2009; Hirsch-Kreinsen et al., 2006; von Tunzelmann & Acha, 2005). In the OECD working paper Hatzichronoglou (1996, p. 4) states that "Firms which are technology-intensive innovate more, win new markets, use available resources more productively and generally offer higher remuneration to the people that they employ. High technology industries are those expanding most strongly in international trade and their dynamism helps to improve performance in other sectors (spillover)."

This view has led to a tendency to understate and underestimate the importance of technological innovation outside R&D intensive fields. On a sample of 1663 German firms, using firm-level data, Kirner et al. (2009) showed that the high-, medium- and low-tech sectors are each comprised of a considerable mix of high-, medium- and low-tech firms. Only about half of the firms from all three sectors matched that classification when measured by R&D intensity at firm level. This finding clearly implies that due to high intra-sectoral heterogeneity the effects of R&D intensity on innovation performance need to be analyzed at the firm level. Thus, generalized statements about sectors with regard to the link between R&D intensity and innovativeness are limited by intra-sectoral heterogeneity.

Low- and medium-tech industries are often viewed as old-fashioned since, compared to high-tech industries, their markets are often relatively mature, slow-growing, and subject to both over-capacity and high levels of price competition. Nevertheless, this does not automatically mean that their products and processes cannot be highly complex and capital intensive (Robertson et al., 2009). Furthermore, when compared in terms of output, capital invested or employment, low- and medium-tech industries are predominant in the economies of both highly developed and developing countries. They account for more than 90% of output in the EU, USA and Japan (Robertson & Patel, 2007; Hirsch-Kreinsen et al., 2006; Sandven et al., 2005; von Tunzelmann & Acha, 2005). As Sandven et al. (2005) note, their contribution to aggregate growth is likely to outweigh that of high technology sectors. Indeed, if low- and medium-tech industries were in fact non-innovative, with attendant decreasing productivity levels, it would consequently result in decreasing levels of national GDP.

What is also of crucial importance to point out is that none of the sectors can be looked at in isolation as their interaction is what drives both growth and development. That is to say, outputs of high-tech sectors are only of value when used together with outputs of other, less technology-intensive, industries. Conversely, low- and medium-tech firms are often major customers of high-tech innovators. Although firms from low- and medium-tech sectors invest less in R&D measured as a percentage of revenues, and are also less innovative, they are nevertheless actively engaged in developing new products and, in particular, new production processes (Robertson & Patel, 2007; Kirner et al., 2009).

Firms in low-tech industries also appear to have the ability to continuously innovate process designs, which results in their value-added processes being of higher quality compared to medium- or high-tech companies. It appears that they compete in terms of the quality of their production processes, which consequently enables them to differentiate themselves from their global competitors via the excellence of their product quality and reasonable process costs (Kirner et al., 2009).

3.2 Specifics of service innovation

Service innovation appears to be in accordance with the previously above mentioned Schumpetrian definition of innovation as service innovations do create new possibilities for further added value, and also stretch beyond the mere technological product and process innovation. Moreover, studies also confirm that services can be, and indeed are, innovative (Coombs & Miles, 2000). Nonetheless, the vast majority of research on innovation chooses to focus on the manufacturing sector.

In order to shed some light on what the specifics limiting the research of service innovation may be, it is best to explore some of the established concepts in relation to service innovation. For a more detailed overview of issues arising in service innovation see Drejer (2004). One of these specifics is the so called "ad hoc innovation". According to Gallouj and Weinstein (1997, p. 549) it describes an "interactive (social) construction to a particular problem posed by a given client", and is a concept with which mostly deal consultancy services. Although ad hoc innovation does not admit of direct reproduction, it can be reproduced indirectly through codification and formalization (Sundbo & Gallouj, 2000). "External relationship innovation" is defined as the particular relationships a firm establishes with its partners (customers, suppliers, public authorities or competitors) (Djellal & Gallouj, 2001) and can be characterized as a subset of organizational innovation. The issue that arises with organizational innovation is that due to it being highly firm specific, it is difficult to formulate it at an aggregate level admitting of comparative analysis (Storey & Easingwood, 1998; Boyt & Harvey, 1997; OECD/Eurostat, 1997, p. 43). Steps have also been taken to theoretically standardize services (Tether et al., 2001), yet despite progress definitions are still not all-encompassing and therefore not generally applicable.

However, the importance of the service sector in national economies should not be overlooked. In terms of employment in the non-financial business sector, services were the largest sector in the 27 EU member states in 2005, accounting for 60%, ahead of industry at 29% and construction at 11%. In member states Latvia, Malta, Estonia, Romania, Luxemburg, Slovenia and Cyprus the weight of value added as the percentage of total value added of the non-financial business economy exceeded that of employment, which indicates relatively high labour productivity in services. Between 2000 and 2004 employment in the services sector in EU-25 grew by 12%, while the growth rates of the non-financial business economy as a whole and that of the employment rate in industry were 6% and -5% respectively (Alajääsko, 2008, p. 1-4).

It remains an open question as to the extent to which parallels can be drawn between manufacturing and services when defining and studying innovation. Coombs and Miles (2000) list three different approaches:

- an assimilation approach which treats services as being similar to manufacturing,
- a demarcation approach claiming that service innovation is distinctively different from innovation in manufacturing, requiring new theories and instruments, and
- a synthesis approach suggesting that service innovation brings to light elements of innovation hitherto ignored, which are relevant for both manufacturing and services.

Given the large body of research on innovation in manufacturing, the assimilation approach provides the most background knowledge on which to build. Studies following this approach make use of subordinate surveys which apply to services definitions as well as questionnaires that were originally developed for manufacturing activities. It is argued that the technology-focused perception of innovation is too narrow to enable a thorough understanding of the dynamics in either services or manufacturing (Drejer, 2004, p. 554). Nevertheless, several parallels have been established. Sirilli and Evangelista (1998) observe more similarities than differences between services and manufacturing with respect to a range of basic dimensions of innovation processes, namely; the propensity to innovate, sources of information, objectives of innovation, and obstacles. Hughes and Wood (1999) further conclude from a sample of 576 small- and medium-sized manufacturing and service firms that differences within each of the two sectors are in fact greater than those between them.

3.3 Innovative performance

Innovative performance – unlike innovation performance, which is considered a separate indicator and measure of the economic success of innovation – refers to new-product development in a broader sense (Hagedoorn & Cloodt, 2007; Marsili & Salter, 2006; Ahuja & Katila, 2001; Lanjouw & Schankerman, 1999). A separate measure is also innovativeness which is defined as the capacity to introduce some new process, product or idea in a given organization (Hult et al., 2004).

Product innovation is recognized as a key condition of business success (Chapman & Hyland, 2004). A successful new product development process contributes to the financial success of the product, and consequently to the overall business success of a firm via two paths (Brown & Eisenhardt, 1995). A productive process lowers costs and hence makes lower and more competitive prices possible. A faster process further ensures strategic flexibility and shorter lead times. Product effectiveness, on the other hand, is demonstrated through various product characteristics, among them low cost, unique benefits and fit-with-firm competencies. Products endowed with these characteristics are also more appealing to consumers (Zirger & Maidique, 1990).

Indicators of innovative performance to be found in literature include: R&D inputs- usually R&D expenditure, including past R&D expenditure-, patent counts, new-product announcement and aggregated constructs of these indicators. Different sectors are characterised by different levels of both innovation inputs and innovation outputs (Tidd et al., 1996), which makes cross-industry comparisons problematic. Attempts have been made to account for this variation by applying the variable of "technological opportunity", which it is difficult to measure and model (Klevorick et al., 1995).

R&D intensity alone does not necessarily reflect innovative intensity for several reasons. Firms tend to broaden their base of R&D expenditure with the purpose of eventually taking advantage of possible tax cuts when such innovation policies are in place. At the same time, R&D represents only a fraction of innovation, a fact which holds especially for companies with less formalized R&D functions. Consequently, total R&D expenditure is difficult to define (Bougrain & Haudeville, 2002). Authors Cassiman, Veugelers (2006) and He, Wong (2004) measure innovative performance as a share of sales, which consists of improved products and new generation products. A separate stream of research employs design as a measure of innovation and design awards as indicators (Hertenstein et al., 2005; Gemser & Leenders, 2001).

Patent counts and new product announcements are also biased measures as differences in the propensity of firms to patent or publish will inevitably affect such measurements (Frumau, 1992). One additional factor that negatively affects the propensity of firms to obtain patents is the considerable cost of registering a patent and the complexity of the procedure. Indeed, not only is it time-consuming but firms also need to disclose many technical details. The speed of technological progress in some industries renders patents obsolete, especially for smaller firms with fewer resources. A significant number of firms also find the protection offered by patents to be insufficient, the exception being the pharmaceutical industry (Mazzoleni & Nelson, 1998; Mansfield, 1984, p. 145).

A noteworthy finding emerged from the Booz & Company management consultancy firm's 2008 compilation of its fourth annual ranking of the world's leading firms according to their investment in R&D (Jaruzelski & Dehoff, 2008). Their report, titled 'Global Innovation 1000', showed no evidence that there has been a link thus far between a firm's investment in R&D and improved financial performance.

3.3.1 Incremental and radical innovation

Product/service innovation can refer to any change in features or design as such, these changes being either incremental or radical. Radical innovations are innovations that are new to either the firm, market, or industry. It is "a product, process, or service with either unprecedented performance features or familiar features that offer significant improvements in performance or cost that transform existing markets or create new ones" (Leifer et al., 2001). These innovations are typically characterized by the incorporation of a substantially different and new technology, providing higher customer benefits compared to products already available. Incremental innovations refer to adaptations, refinements, enhancements or line extension by adding new features and thus offering additional benefits. If incremental innovations incorporate changes in the underlying technology, these tend to be small and place only limited strains on a firm's existing competencies (Garcia & Calantone, 2002). For companies to remain competitive in the short term, incremental innovation can be a good source of competitiveness. However, long-term growth is linked more closely to radical innovation (Morone, 1993, p. 220). In order to spread resources strategically, companies should actively pursue both strategies- incremental and radical innovation- simultaneously.

Relative to radical innovations, incremental innovations are more market-driven and based on market analysis; therefore, they are more likely to be successfully commercialized and less likely to suffer from insufficient demand, an advantage not shared by radical innovation. The lower profit potential of incremental innovation is, on the other hand, offset by the high probability of technical completion (Varadarajan, 2009; Ali et al., 1993). Kanter (2006) observes that successful innovators can be viewed as an innovation pyramid consisting of a few substantial risks at the top, a larger number of promising midrange ideas in test stage and a broad base of ideas at an early stage of development. Even though incremental innovations as competitive differentiation advantages of a firm are at risk of being neutralized by competitors' actions and may yield only marginal gains, their cumulative effect can still be expected to be significant.

Varadarjan (2008, p. 2) lists the following roles of incremental innovation in the competitive strategy of a firm:

- extending the time horizon of the revenue stream from radical innovations,
- entering new markets in product categories in which the firm currently has a presence (new types of markets – e.g. entering the business-to-business (B2B) market from the business-to-consumer (B2C) market; new market segments; new geographic markets),
- entering new product-markets in product categories in which the firm currently does not currently have a presence (new product-markets that are pres-

ently fragmented industries; new product markets that emerge or become attractive as a consequence of changes in the legal and regulatory environment; related new product-markets with entrenched competitors),
- achieving and defending product category leadership by product differentiations that enable a firm to pursue a multi-brand strategy through differentiated product positioning and target marketing (pre-empting shelf space by pre-empting potential entry points of competitors; responding to price sensitivity and variety-seeking, behaviour driven brand switching; protecting flagship brands with flanker brands),
- enabling the firm to command a higher price relative to the product being superseded by the incremental innovation, or a price premium relative to competitors' offerings, in order to achieve higher margins, and
- adapting to the structural constraints of the industry ecosystem.

When incremental innovations are used for line extension, such a product proliferation strategy can increase the overall demand for a firm's products, affect supply by increasing costs and deter competitors from entering, thus allowing the incumbent firm to increase prices (Bayus & Putsis, 1999). Those incremental innovations that appear in the form of additional new features in a firm's existing product (range) provide positive differentiation by giving a product perceived advantages over the competition. In the eyes of consumers, brands with a greater number of features rank higher in their choice set (Brown & Carpenter, 2000).

Koen and Kohli (1998) developed a survey on a sample of large companies with the aim of evaluating the source of ideas for new products which had been commercialized for at least 5 years. They analyzed 3 types of products; radical products, platform products and incremental products. A radical product is one which provides the customer with completely new benefits. A platform product provides a large number of improvements and involves a significant change, while an incremental product involves only minor changes in the offering. Ideas for new radical products come from the cooperation of the engineer/scientist and the customer. This is neither technology push nor marketing pull. Customer needs for radical products are tacit and the customers have difficulty expressing their needs beyond the obvious. Similarly, the engineer/scientist does not understand how the new technologies can fulfil the future needs of the market place. This data suggests that the technologist and customer must liaise on a solution so both parties understand how new technology can be used to fulfil unexpressed customer needs.

In contrast to a radical innovation product there is no direct customer involvement in platform and incremental products. For platform products the engineer/scientist still plays the most important role, though the division president and sales manager are also involved. This data suggest that ideas for new platform products come from the technologist and the customer knowledge residing within the company. The division president and the senior sales manager typically have in-depth knowledge of their customers' expressed needs and wants and can accurately describe them to the technologist.

The engineering scientist no longer plays a key role in idea development in incremental product development where the ideas come instead from various different sources. Although the legitimacy of generalizations is limited by the sample size of the data set, the findings suggest that new incremental product direction is clear to the innovator since both the customer needs and the technologies are well understood.

With respect to types of cooperation in R&D activities, Tödtling et al. (2009) confirm through their work on a sample of Austrian firms that more advanced innovations require a higher degree of internal R&D and patenting. These innovations are further supported by cooperation with universities and research organizations. As is clear from their very name, they rely more on scientific inputs than less advanced innovations. The introduction of incremental innovations also requires some amount of R&D-activity, but relatively less, as in such cases cooperation with service firms that supply practical knowledge is of much greater importance. The authors also find that less binding forms of knowledge interaction, such as information exchange, have no influence on innovative activity.

3.3.2 Technological and market turbulence

Greenly and Oktemgil (1997) suggest that as a moderating effect, the external business environment may severely influence managerial choice. Increasing environmental turbulence shortens the life span of many resources (Grant, 2001, p. 13), hence managers are expected to formulate strategies in accordance with the relevant information about the environment. It is argued that successful new product development depends strongly on the characteristics of the competitive environment in which the industrial firm operates (Langerak et al., 1997); more specifically, technological and market turbulence (Calantone et al., 2003).

How managers perceive the environment will also be reflected in their actions and the innovative strategy they choose to pursue. It is important that firms recognize environmental changes and adapt accordingly (Leonard-Barton, 1992). Technological and market turbulence are those two moderating effects that influence new product development strategy planning (Calantone et al., 2003).

Technological turbulence refers to the perception of whether a firm is able to predict accurately and understand thoroughly specific aspects of the technological environment. Technological and complementary competencies are key to addressing changes and achieving superior performance in environments with high technological turbulence (Wang et al., 2004). Wheelwright and Clark (1992, p. 99) place special emphasis on the state of industry maturity, claiming that in relatively young industries every developmental effort appears to be aimed at broadening the firm's market coverage, while the incremental changes are targeted primarily at correcting deficiencies in the underlying platform products.

Market turbulence, on the other hand, reflects rapidly changing buyer preferences, wide-ranging needs and wants, competition intensity and an ongoing emphasis on offering new products (Hult et al., 2004). Firms operating in high market turbulence therefore tend to constantly produce innovations in order to respond to both the changes in demand and the presence of strong competition. They need to develop superior marketing competencies together with strong complementary competencies.

3.4 Lisbon strategy and innovative activity in the European Union

In 2002, the EU Member States set in motion a new strategy based on economic reforms, the purpose of which was to enhance the competitiveness of the region. The so-called Lisbon Strategy attempts to achieve, through various measures, the following objectives (Kok, 2004):

- a greater amount of R&D and innovation,
- a more dynamic business environment,
- increased investment in people, and
- the greening-up of the economy.

One of the initial overall objectives with respect to R&D and innovation was to raise the overall research investment in the EU from 1.9% of GDP to 3% by 2010. Upon the realization that the interim results of the strategy were rather modest, the Lisbon strategy was simplified in 2005. The microeconomic guidelines adopted are largely - either directly or indirectly- related to R&D and innovation as competition, investment and innovation are expected to contribute to job creation and long-run growth. Consequently, national and regional programmes for the period 2007-2013 are increasingly targeted at investments in knowledge and enhancing the innovation capacity (Commission of the EC, 2005).

According to data for the year 2006, the EU is spending about 1.85% of GDP on R&D (Commission of the EC, 2007). The share of R&D expenditure ranges across Member States from below 0.5 % to nearly 4 % of GDP. Compared to data for 2000, the level of R&D spending has slightly decreased. The challenge that remains is to develop economic framework conditions, instruments and incentives conducive to companies investing more in R&D. Economic framework conditions encompass smoothly functioning financial and product markets and also the efficient enforcement of intellectual property rights. In order to support innovative activity, the proposed innovation strategy is set rather broadly along these lines, addressing:

- intellectual property rights,
- standardisation,
- the use of public procurement to stimulate innovation,
- joint technology initiatives,
- boosting innovation in lead markets,
- encouraging cooperation between higher education, research and business,
- encouraging innovation in regions, innovation in services and non-technological innovation, and
- improving businesses' access to risk capital.

For the purpose of measuring the innovation indicators and providing assessments of national innovation performance for the Member States, two main instruments are in place; namely the Community Innovation Survey (CIS), and the European Innovation Scoreboards (EIS), the second being heavily reliant on data obtained by Eurostat and CIS (OECD, 2006).

The most recent CIS survey for which the data is readily available is that of 2005, the fourth such survey carried out in consecutive years. The observation period was from 2002 to 2004. The questionnaire used is based on the 1997 Oslo Manual and focuses on:

- product, process, ongoing and abandoned innovation,
- innovation activity and expenditure,
- intramural research and experimental development (R&D),
- effects of innovation,
- public funding of innovation,
- innovation co-operation,
- sources of information for innovation,
- hampered innovation activity,
- patents and other protection methods, and
- organizational and marketing innovations in the enterprise.

Questions referring to activities and effects are evaluated by respondents according to their importance. The scale used is a four-point scale with categories "high", "medium", "low" and "none/not used". Included in the target population are all firms with more than 10 employees from the following sectors: industry, wholesale trade, transport, storage and communication, financial intermediation, computer and related activities, architectural and engineering activities and technical testing and analysis.

The results of the fourth CIS (Eurostat, 2007) show that in the EU-27 42% of firms reported some form of innovation activity. The highest proportion of companies manifesting innovation activity was observed in Germany (65% of total firms), Austria (53%), Denmark, Ireland and Luxembourg (52% each), Belgium (51%) and Sweden (50%). Conversely, the lowest rates were reported in Bulgaria (16%), Latvia (18%), Romania (20%), Hungary and Malta (both 21%). The share of enterprises with innovation activity for Slovenia was 27%. Slovenia fared a lot better with regard to innovation co-operation. While 26% of all innovative firms took part in innovation cooperation, Slovenia placed second at 47%, behind Lithuania (56%) and ahead of Finland (44%). The lowest levels were reported in Italy (13%) and Germany (16%).

Unlike CIS, EIS calculates a Summary Innovation Index of innovation performance, based on 26 indicators. Indices are composed for European countries as well as Japan and the USA. EIS was first used in 2000 as a direct consequence of the adoption of the Lisbon strategy.

Based on the innovation performance results of the EIS 2007, the countries have been divided into the following groups (European Innovation Scoreboard, 2008, p. 7):

- innovation leaders: Denmark, Finland, Germany, Israel, Japan, Sweden, Switzerland, the UK and the US,
- innovation followers: Austria, Belgium, Canada, France, Iceland, Ireland, Luxembourg and the Netherlands,
- moderate innovators: Australia, Cyprus, Czech Republic, Estonia, Italy, Norway, Slovenia and Spain, and
- catching-up countries: Bulgaria, Croatia, Greece, Hungary, Latvia, Lithuania, Malta, Poland, Portugal, Romania and Slovakia.

Sweden is the most innovative country of those deemed innovation leaders, which can be attributed largely to strong innovation inputs despite its lower efficiency relative to certain other countries when transformation of these inputs into innovation outputs is in question. The above groups, however, seem to have been relatively stable over the last five years. There have been changes in the relative ranking of countries within groups but this does not appear to extend to changes between groups. At this point, only Luxembourg is on the verge of entering the group of innovation leaders.

3.4.1 Community Innovation Survey – Slovenia

As previously mentioned, the results of the fourth CIS for Slovenia show that only 27% of Slovenian firms from selected industrial and service sectors engaged in innovation activities during the period 2002-2004 (Celikel-Esser et al., 2007). Products new to firms made up 14% of the total turnover, while this number falls to 7 % for products entirely new to the market. Half of the enterprises attributed perceived "improved quality in goods and services" to innovation and its direct results. 32% of the innovative firms "entered in a new market or increased their market share" during the observed period. A comparable share (38%) of firms "increased the range of good and services" (38%). Almost a third (31%) of innovators reported that innovations led to an "increased capacity of innovation or service production" as well as "improved flexibility of production or service provision". As a result of innovation, almost 28% of the enterprises in question were able to "reduce labour cost per unit of output", 19% reduced their environmental impact and 17% succeeded in cutting materials and energy per unit of output. Indicators on intellectual property rights and registered trademarks are not available (Figure 1).

Firm size appears to have a strong effect on innovation activity in Slovenia as large firms innovate significantly more (70 % of all large firms) than medium-sized (41 %) and small (19 %) firms. This finding is consistent with the data on firms having introduced new products to the market. Large firms lead with 20%, followed by medium sized firms (12%) and small firms (4%). The majority of large innovators (66%) engage in innovation cooperation, while figures for medium sized firms and small firms are 52% and 38% respectively. Smaller disparities due to size can be observed for the share of enterprises that increased the capacity of production and service provision (35% for large enterprises and roughly 31% for small and medium sized firms). Manufacturing witnesses more than double the amount of innovation (35 % of firms) compared to the service sector (16 %).

Figure 1: Ranking of Slovenia among EU-27
according to selected CIS innovation measures

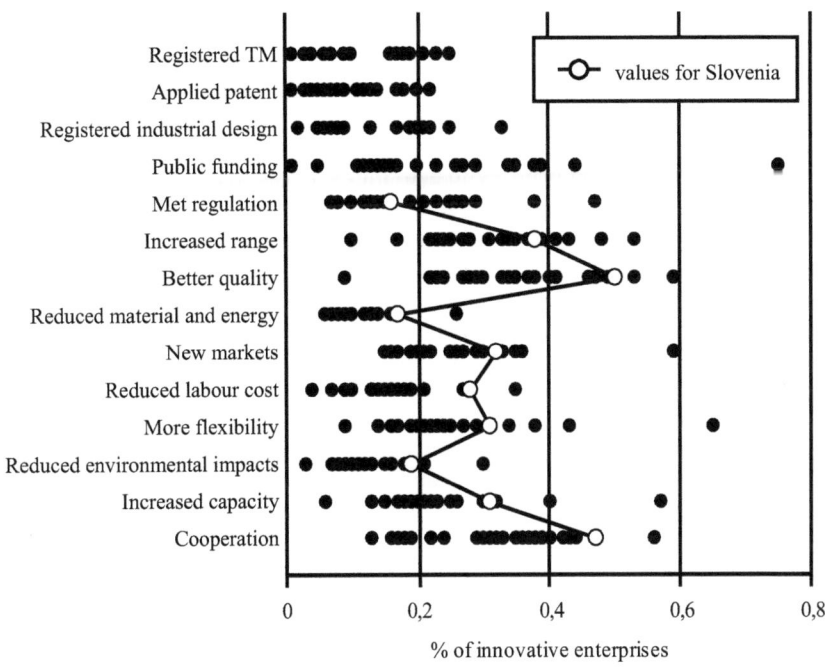

Source: Celikel-Esser et al., 2007.

3.4.2 Technology leaders and followers

Innovation and corresponding competencies demonstrate some specific characteristics when a distinction is made between firms that are technology leaders and those that are technology followers. Forbes and Wield (2000) state that basic research and applicative research enable technologically advanced companies – technology leaders – to create new knowledge and to promote new technologies. Followers, on the other hand, develop indigenous technology learning capacity or, in other words, the abilities to use existing technological solutions in a more efficient manner. It is therefore characteristic of technologically advanced companies to introduce new products, which are new for the market, by using new technologies and by transforming existing technological solutions into new ideas. Being a technology leader demands substantial investments that contain large elements of risk due to the high likelihood of failure. Holding a leading position in innovation also requires the establishing and maintaining of close relations with key sources of relevant, new knowledge as well as with the needs and responses of customers (Porter, 1998, p.165-166). Followers tend to rely more on incremental than on radical innovation, the former being based on basic and applicative research as well as on industrial design that provides these firms with an opportunity to supply market niches and achieve high value added. By imitating leaders, followers have the opportunity to learn from the experience of technological leaders. However, they need to commit firmly to activities such as competitor analysis and intelligence, reverse engineering, cost cutting and learning in manufacturing. Reverse engineering refers to testing and dissembling of competitors' products to gain an understanding of how they function and what benefits they offer to the customers (Tidd et al., 1997, p. 121). As presented in the chapter on the Lisbon strategy, the Slovenian economy ranks as a moderate innovator according to the EIS study, which could, in a broader context, be described as a group of technology followers.

Metcalfe (1995) describes National Innovation Systems (NIS) as "that set of distinct institutions which jointly and individually contribute to the development and diffusion of new technologies and which provides the framework within which governments form and implement policies to influence the innovation process. As such it is a system of interconnected institutions to create, store and transfer the knowledge, skills and artefacts which define new technologies". Authors Nelson and Rosenberg (1993) state that there appears to be a strong belief that the technological capabilities of a nation's firms are a key source of their competitive process and of national dimension. The latter implies they can be built by national action. For a comparative overview of existing methodologies on measuring technological capabilities at the country level, see (Archibugi & Coco, 2005). The country rankings are based on aggregate measures which cover areas such as generation of technology and innovation (most often measured by patents), infrastructure and technology diffusion, human capital and competitiveness.

Even though public intervention is generally encouraged to promote technical advancement, differences among nations call for customised approaches to industrial development. Case studies of NIS (Nelson, 1993) point at their five main features:

- education and training,
- science and technology capabilities,
- governance/business balance,
- industrial structure, and
- interactions among the different parts of the innovative systems.

In education and training the main differences arise from the number of students enrolled in different levels of education and the scientific disciplines students choose to pursue. Science and technology capabilities or, in other words, the level of resources devoted by each country to formal R&D and other innovation-related activities (among them design, engineering, tooling-up) is a basic characteristic of NIS. The vast majority of the world's R&D activities are carried out in industrially advanced countries. Even among the OECD countries, significant differences in R&D intensity are witnessed. Formal R&D activities can be found at the core of NIS only in a small number of countries, among them the USA, Japan, Germany, Switzerland and Sweden.

Since firms act as the principal agents of technological innovation, innovative activities on a national level are to a great extent under the influence of national industrial structure. Large firms are more likely to commit to long term investment plans and basic research programmes. The level of competition companies face in their domestic market is also a decisive factor in determining their R&D investment choices.

Case studies recognize the level of coordination among different players as the most important driver of technological change via NIS. In some countries this means interaction between government and national champions or between government and industry in general. One example of small firms developing a common competitive strategy can be found in the activities of Italian industrial districts (Malerba, 1993). Oftentimes these interactions result in an improved diffusion of innovation and the multiplication of its effects. On the other hand, failure to do so can impede the economic effectiveness of the resources invested in science and technology.

While some of the key characteristics of NIS can be transferred among countries, others cannot. The manner in which a country should approach the construction of its technological competence is highly path-dependent. There is not just one single model of an innovative system that can lead to industrial development. Even heavy investments in industrial R&D and technology have, historically, not been proven to be a necessary factor. Nevertheless, the potential of innovation systems should not be neglected. New and more effective forms of technological expertise have given rise to world leaders. A new innovation system provides a nation with an advantage over competitors and can become the driving force of subsequent economic superiority. Technology follower nations can choose from various methods regarding how to organize their innovative system; however, there is much less freedom for those competing for the leading position. The organization of industries in a national economy tends to be technology specific, while the impact of innovation is to a large extent influenced by the overall national economic activity (Nelson, 1993, p. 518). Nevertheless, countries should refrain from supporting national champions alone and rather create policies for improving wider infrastructures on a national level (Reich, 1991, p. 135)

Archibugi and Pietrobelli (2003, p. 880) provide advice to developing countries on how to maximise the benefits of the globalization of technology. The importing of foreign technology has, as such, a negligible learning impact unless it is accompanied by local policies to promote learning, human capital and technological capabilities. Public policies should thus focus primarily on motivating foreign firms to move from: (a) exporting their products to (b) producing locally, and transferring a technological component. Additionally, it is often more advantageous for a developing country to set up inter-firm strategic technological agreements than simply play host to the production facilities of foreign firms. Public policies should, therefore, also aim to "upgrade" FDI to strategic technological partnering. Collaborations among public and business organizations can also be of considerable benefit to developing nations. Therefore, policies at both the national and intergovernmental levels should consider these collaborations as a channel of choice for transferring and acquiring technological competencies.

4 Theories of competitive advantage

Industrial organization economics emphasize industry attractiveness as the primary basis on which superior profitability is founded. The steps this requires of strategic management range from seeking favourable industry environments, and locating attractive segments and strategic groups within industries, to moderating competitive pressures by influencing industry structure and competitors' behaviour. However, empirical research fails to support the link between industry structure and profitability (Grant, 2001). What studies do imply is that differences in profitability within industries are of greater significance than differences between industries (Schmalensee, 1988). In other words, competitive advantage takes precedence over external environments when accounting for inter-firm profit differentials between firms. In this respect, three views have emerged which attempt to explain the sources of a firm's competitive advantage, namely: the resource-based theory, dynamic capabilities theory and competence-based theory.

In the following chapters are presented these three theories of competitive advantage. The emphasis is on the competence based theory, which are linked to new product development activities with the aim of showing in what way firms can build competitive advantage via R&D and innovative activities. A comparative summary is provided in Appendix A.

4.1 Resource-based theory

The resource-based theory of competitive advantage was developed due to increased interest in the role of a firm's resources as the foundation of firm strategy. At the same time, it reflects dissatisfaction with the static, equilibrium-based framework of industrial organization economics. Its contribution is twofold and includes both the corporate strategy level and the business strategy level. At the corporate strategy level the attention was focused on the role of firm's resources in determining the industrial and geographical boundaries of the firm's activities (Grant, 2001; Teece, 1980). Simultaneously, at the business strategy level arose, among others, analysis of competitive imitation (Rumelt, 1984; DeFillippi, 1990), the appropriability of returns of innovation (Teece, 1988), the role of imperfect information in creating profitability differences between competing firms (Barney, 1986), and the means by which the process of resource accumulation can sustain competitive advantage (Dierickx & Cool, 1989).

Penrose (Penrose & Pitelis, 2009, p. 39-57) studied how a firm's internal management processes affected its behaviour with respect to why and how firms grow. She viewed firms as a collection of productive resources and suggested three roles of management that limited a firm's growth:

- management failing to recognize opportunities in market demand that could be provided for by the available resource,
- the extent of management's ability to combine existing resources with new ones required for entering new geographic or product markets and
- the willingness of management to take relevant risks arising from the desire to serve new market demands.

Wernerfelt (1984) introduced the concept of resource position barriers, a theoretical tool which refers to barriers inflicted by higher costs related to new resource adoption. The first movers in creating and using a given resource, be it made up of tangible or intangible assets, enjoy lower costs compared to those acquiring an existing resource. The underlying explanation for this is the advantage created by having experience with the resource. Thus, resources that are subject to the experience curve are regarded as attractive since they can lead to considerable profit. Wernerfelt also highlighted mergers and acquisitions as a way for firms to acquire bundles of attractive resources in highly imperfect resource markets.

Barney (1986, 1991) connects the concept of firm resources with the sources of sustained competitive advantage. He used a very broad definition of firm resources, namely as all assets, capabilities, organizational processes, firm attributes, information, knowledge, etc. under a firm's control that facilitate strategies aimed at improving the firm's efficiency and effectiveness, resulting in the earning of economic profits. He classified resources further, subdividing them into three categories: physical capital resources, human capital resources and organizational capital resources. A sustained competitive advantage is achieved when a firm employs resources within a value creating strategy that cannot be adopted by current or potential competitors. Therefore, only heterogeneous and imperfectly mobile resources can take on this role. Barney went on to propose four additional requirements for these resources:

- they must be valuable either in virtue of being used for exploiting opportunities or as a way to neutralize threats,
- the resource must be rare and not possessed by a large number of existing or potential competitors,
- they must be imperfectly imitable, and
- there must not be substitutes of equal qualities.

Few resources are productive on their own. What it takes for resources to be a part of a productive activity is the cooperation and coordination of resource teams. The capacity of a team of resources to perform some task or activity is termed a capability. Resources are considered to be the source of a firm's capabilities, which are, in turn, the main source of competitive advantage. The resource based approach to strategy is concerned not only with the deployment of current resources, but also with the ongoing development of the firm's resource base.

How sustainable the competitive advantage will be depends on the durability, transparency, transferability, and replicabiltiy of resources and capabilities. Capabilities as such are possibly more durable than individual resources. The reason for this is that capabilities can be maintained intact despite individual resources being replaced along the way. The complexity of capabilities is particularly relevant to the sustainability of competitive advantage. Simply put, the larger the number of diverse resources that together constitute a capability, the higher its degree of complexity. Imperfect transferability makes it difficult for other firms to acquire the desired resources or capabilities and imitate success. Highly complex organizational routines affect the transferability of capabilities in the same way (Grant, 2001).

The premise from which this view stems is that when formulating a strategy, firms begin by carrying out a revision of their mission statement regarding their identity and purpose. This helps them answer questions pertaining to what the firm's business is and which markets they serve, who the customers are and what customer needs they aim to satisfy. In a volatile environment with constantly changing customer preferences, an externally focused orientation is not a stable basis for long-term strategy. In this respect, a firm's own resources and capabilities provide a more solid ground for defining the firm's identity.

This view is often criticized on the grounds that different combinations of capabilities might generate the same value and therefore do not represent competitive advantage (Priem & Butler, 2001).

4.2 Dynamic capabilities theory

The dynamic capabilities theory is, at its core, an extension of the resource based view and the role of resources. It developed from a growing awareness of the importance of a firm's relative abilities to: use current resources, create new resources, and devise new ways of using current new resources (Sanchez, 2002, p. 150).

Nelson and Winter (1982, p. 73, 124) looked at how firms innovated and induced changes in economic activity. They presented organizational routines as those repetitive activities that a firm develops in its use of specific resources. In order to explain the role and position of routines in an organization, an analogy can be made with skills and what these are and mean to an individual. New skills are developed by improving existing skills and the same holds for routines. Teece et al. (1997) introduced the notion of dynamic capabilities as a firm's ability to "integrate, build and reconfigure" internal and external routines. They drew attention to path dependencies, which constrain a firm's ability to make short-term adjustments to existing routines, to develop new ones, and to imitate those of competitors. Path dependencies are created by organizational and managerial processes as well as a firm's current resource position.

Amit and Schoemaker (1993, p. 36) combined the concepts of resources and dynamic capabilities. With the term "strategic assets" they refer to "the set of difficult to trade and imitate, scarce, appropriable, and specialized resources and capabilities that bestow a firm's competitive advantage". Certain strategic assets will be subject to market failures and will this way become the "prime determinants of organizational rents" in an industry. Organizational rents in fact refer to economic rents that can be captured by the organization rather than the owner of the resources and capabilities it confers and uses. The set of these so called strategic industry factors, however, keeps changing and cannot be predicted. The resulting uncertainty leads to complexity and social conflict in managerial processes in dealing with challenges of the future. Therefore, the authors introduced cognitive and social dimensions of the managerial decision-making process.

4.3 Competence-based theory

A consequence of the incorporation the concepts of resources and dynamic capabilities, the competence perspective on strategy emerged in the early 1990s. According to Sanchez (2002, p. 152), it expands on the complex interplay of resources, capabilities, organizational processes, managerial cognitions and social interactions within and between firms. Hamel (1994) put forward several arguments as to why a more integrative theory of strategic management based on the concept of organizational competence would be of interest. They saw in it a potential to obtain new insights into how creating and sustaining competitive advantage depends on a firm's capabilities in managing knowledge resources. An elaborated concept of competencies could provide tools to help firms become more effective at combining resources and capabilities in building and leveraging organizational competencies. Furthermore, such a concept could aid in improving understanding of how firms think and act systematically with respect to creating strategic and operational flexibility. This includes management processes shaping the firm's vision of the future and such better understanding would also help identify, as well as create, new competencies. In the next chapter the competence-based view is presented in more detail, with firm competencies and their role in creating competitive advantage being elaborated on in the following section.

5 Firm competencies and competitive advantage

Despite the growing volume of research and attempts to develop a unified definition of the underlying theoretical concepts of the competence based theory, their use and application in research is still somewhat confusing. This can be seen especially in the use of the terms capabilities and competencies, since many authors fail to make the distinction (Andrews, 1998; Hamel, 1994; Ansoff, 1965, p. 6). Chiesa and Manzini (1997) observed three reasons for various definitions appearing in the literature:

- similar concepts described with different terminology,
- similar terms describe different levels of activities within organizations, and
- many researchers take on a static view of competencies that does not take into account their creation and leveraging.

The existing and prevailing definition of competence will be presented first. Different uses of capabilities and competencies will be further addressed in the section regarding measurement.

Hamel defines competence as a "bundle of constituent skills and technologies, rather than a single discrete skill or technology" (Hamel, 1994, p. 11). The implications of this definition are twofold; firstly, a competence is basically the integration of a variety of individual skills and, consequently, what distinguishes the core competencies of firms is the uniqueness of such integration. To regard a coherent cluster of assets, knowledge and skill as a competence, it must add value to end products, it has to apply to a range of different markets and be difficult to develop and imitate (Prahalad & Hamel, 1990, p. 84). Competencies enable a firm to deliver a fundamental customer benefit that is reflected in characteristics such as reliability or user-friendliness, among others.

The term "core competencies" is used to describe central, strategic capabilities. A core competence does not correspond to an asset in the accounting sense. It is rather an accumulation of learning encompassing both tacit and explicit knowledge. An attempt to list every single competency of potential importance to success in a particular business would yield a very long list. Since it is impossible for senior management to focus on all, the inevitable and reasonable goal is to choose those that are key to competitive success; in other words, core-competencies.

The first step toward producing a unified definition of competencies was made by Sanchez et. al (1996, p. 7-11). The objective of the authors was "to develop a vocabulary that is conceptually adequate, internally consistent and capable of serving as a language for discussing competence-based competition". Assets were defined as "anything tangible or intangible the firm can use in its processes for creating, producing and offering its products (goods or services) to a market". Capabilities were described as: "Repeatable patterns of action in the use of assets to create, produce and/or offer products to a market. They are an important special category of assets that determine the uses of tangible assets and other kinds of intangible assets. They arise from the coordinated activities of groups of people who pool their individual skills in using assets." With these definitions a hierarchy of interrelated concepts is established with assets at the top of the pyramid, followed by capabilities and skills, both individual and team.

Lastly, "competence is the ability to sustain the coordinated deployment of assets in ways that help a firm achieve its goals." In order for a firm's activity in using resources and capabilities to be recognized as a competence, it must fulfil the three conditions of "organization (implicit in the notion of co-ordination), intention (implicit in the notion of deployment) and the potential for goal attainment." Here it is important to point out that with these definitions of "competence" and "capability" it is not implied that these terms are to be used interchangeably as suggested by Hamel (1994, p. 12) in his definition of core competence.

Competence maintenance advocates that merely maintaining a firm's current competencies requires the continuous adaptation of current resources and capabilities to changing environmental conditions. Competence building refers to any process by which a firm achieves qualitative changes in existing assets and capabilities, thus creating new strategic options for future actions relevant to the firm's pursuit of its goals. Competence leveraging, on the other hand, stands for applying a firm's existing competencies to current or new market opportunities. Leveraging does not require qualitative changes in the firm's assets or capabilities but may call for quantitative changes (Sanchez et al. 1996, p. 11).

Capabilities, unlike competencies, are focused and manifest themselves within the activities and processes of a function. Hitt et al. (2005) define a capability as the capacity of a set of resources to integratively perform a task or an activity. A capability thus represents a firm's ability to deploy resources that have been deliberately integrated to achieve a desired end state. Competencies are usually a platform of multiple lines of businesses and/or products within a corporation. They are the most important building blocks of cross-functional business processes. This is described as collectiveness of competencies, and it should be noted that it is this characteristic that provides companies with opportunities to produce new products or enter new markets. The three elements of collectiveness are across-function, across-product and across-business. The across-function element describes the extent to which a capability is an indispensable element of one or more cross-functional processes, while across-product and across-business elements are measures of the extent to which capabilities are shared by various products and business units respectively (Hafeez et al., 2002, p. 31).

Figure 2 shows the architecture of core competencies as proposed by Hafeez et al. (2002, p. 30-31). The resources are inputs to capabilities. Those capabilities that are more crucial to a firm realizing its business objectives are key capabilities. Only those key capabilities that are both relatively unique and common to various business functions, products and business units are likely to form competencies of a company. This last mentioned research also distinguishes competencies and capabilities. Usually, competencies are not based on a single activity but are represented, or constituted, by a network of capabilities (Sanchez et al., 1996). The authors provide two companies, 3M and Canon, as examples, attributing a firm's competence in R&D to the coordination of several capabilities such as research, product development and experimentation. The product development capability of Canon, a world leader in imagining products, is a result of its expertise in fine optics, precision mechanics, and microelectronics. As a rule of thumb for the aggregation level of competencies, Hamel suggests that there are between five and fifteen core competencies for any individual business. A larger number of identified competencies could already include individual skills, whereas a smaller number would describe so-called meta-competencies. An example of a meta competence of a firm could be marketing, their core competence being customer relationship management, containing within it the constituent skill loyalty-building activities. The critical task is to assess capabilities relative to those of competitors'. Core competencies are those competencies that help a company achieve a sustainable competitive advantage. These are competencies that are, by nature, strategically flexible and dynamic.

Figure 2: The architecture of core competencies

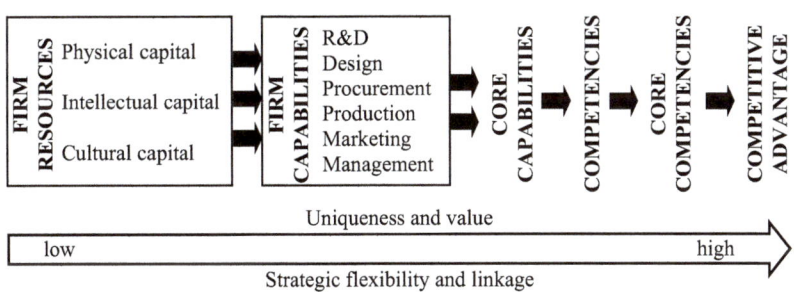

Source: Hafeez et al., 2002.

Changing organization's core competencies is a more time-consuming process than the change of products they themselves make possible. They are created through the "collective learning" of a firm, which comes from the coordination of diverse production skills, integration of different technologies and use of resources and capabilities (Rumelt, 1994, p. xv-xvi).

There are four so-called cornerstones of competence theory embodied in these concepts; more specifically, the dynamic, systemic, cognitive and holistic natures of firm competencies (Sanchez, 2004, p. 519). Dynamic nature refers to the ability of a competence to respond to the dynamic nature of both a firm's external environment and its own internal processes. Inhering in it is sustainability, or the ability to defy changes in market preferences and available technologies as well as to maintain internal organizational dynamics. Loss of internal organizational dynamic results in organizational entropy and can be witnessed as a gradual loss of organizational focus. The loss of focus can be observed as "a narrowing and increasing rigidity in the patterns of activity the organization can or does perform, a progressive lowering of organizational expectation for performance or success, and alike." (Sanchez, 2004, p. 521) Therefore, it is the manager's task to keep providing inputs of energy and attention in order to maintain or even improve the organization's value-creation processes.

A firm's systemic nature of firms refers to the need to coordinate firm-specific assets- those under direct control of the firm- and firm-addressable assets, which lie beyond the boundaries of the firm. Materials and components suppliers, distributors, consultants, financial institutions and customers are examples of firm addressable assets. They can be described as assets that a firm does not own or tightly control but that it can arrange to access and use from time to time (Sanchez et al., 1996).

The third cornerstone – cognitive nature – asserts that competence must include an ability to manage the cognitive processes of a firm in terms of directing organizational assets to specific value-creating activities in an efficient and effective manner. Lastly, holistic nature addresses the multiplicity of individual and institutional interests that interact and are served within any given firm. Therefore, managers must be in a position to define satisfactory organizational goals for all resource providers.

The research in the thesis will adhere to the prevailing definitions described above. The exception will be the distinction drawn between capabilities and competencies in accordance with the definition of Sanchez et al. (1996). To summarize, capabilities will refer to organizational routines and processes, while core competencies are understood as the combination of resources and capabilities that serve as a source of competitive advantage. Capability being developed and combined with other resources therefore becomes a competence. If a competence becomes a building block of the competitive advantage of a firm, it is regarded as a core competence. This definition, as also used by Prajogo et al. (2008), not only provides a more rigorous understanding of the concepts but at the same time offers more precise systematic tools for the purposes of analysis.

5.1 Competencies as a source of competitive advantage

The view of competitive advantage as a function of inherent industry attractiveness and the market positioning of individual firms is most known for the contributions made by Porter and his concept of "the five competitive forces that determine industry profitability" (1998a, p. 4-10). It is a traditional view that helps identify which firm-competencies management should concentrate on. Empirical studies show that industry factors are not the key determinant of the profitability of an individual firm. The direct industry effect has been estimated as being between 16% and 19% of the total variations in profit between business units (Rumelt 1991, Schmalensee 1985). There are two ways in which competencies, by acting as a catalyst in the process of asset accumulation and thus improving it, contribute to the competitive advantage of a firm (Figure 3) (Verdin & Williamson, 1991). One is the deployment of an appropriate set of core competencies across business units within the firm. This, in effect, reduces costs and increases the speed with which new, non-tradable and industry specific assets can be accumulated. Through core competencies a firm may, in this fashion, quickly achieve a desirable position within a new market. On the other hand, core competencies may also allow a firm to maintain or extend its competitive advantage by making it possible for the firm to augment its non-tradable, industry-specific assets more quickly than its competitors. Strategic flexibility is especially essential in markets that are witnessing, or are subject to, significant change for it enables the firm to adapt to changing circumstances. It depends jointly on the firm's resource flexibilities and the co-ordination flexibilities of the firm's managers in coming up with new configurations and uses for both current and new resources (Sanchez, 1995).

Figure 3: A "production function" for competitive advantage

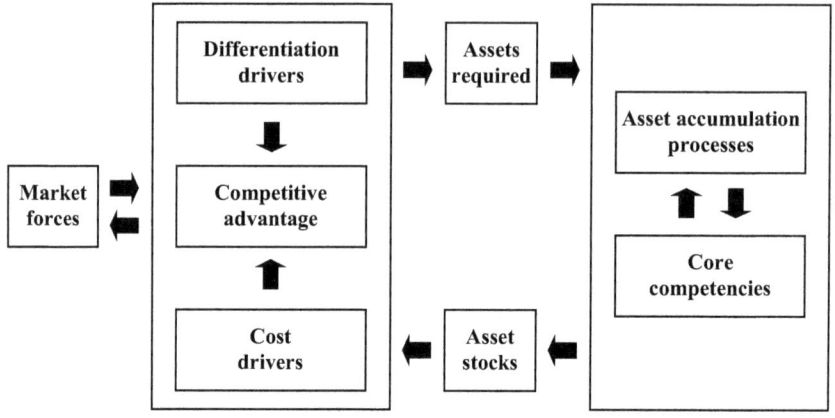

Source: Verdin & Williamson, 1991.

Five main groups of assets addressed within this view are stated below while a more detailed list can be found in Verdin and Williamson (1991):

- input assets – e.g. input assets, loyalty of suppliers, financial capacity,
- process assets – e.g. proprietary technology, functional experience, organizational systems,
- channel assets – e.g. channel access, distributor loyalty, pipeline stock,
- customer assets – e.g. customer loyalty, brand recognition, installed base, and
- market knowledge assets – accumulated information as well as the systems and processes to access new information on the goals and behaviour of competitors, the reactions of customers, suppliers and competitors to different phases of the business cycle.

It is this portfolio of assets on which Porter's various cost and differentiation drivers depend. Examples of cost drivers are economies of scale, learning and spillovers, linkages, interrelationships, integration, timing, discretionary policies, location an institutional factors. Differentiation factors are discretionary policies with emphasis placed on quality and service, linkages, timing, location, interrelationships, learning and spillovers, integration, scale, and institutional factors.

The role of a catalyst in the process of asset accumulation is valuable due to the four following factors, all of which present an obstacle to cheap and rapid asset acquisition: time compression diseconomies, asset mass efficiencies, asset interconnectedness and causal ambiguity (Dierickx & Cool, 1989). Time compression diseconomies appear in connection with the extra cost of accumulating required assets under time pressure. Asset mass efficiencies describe costly accumulations of assets of which the existing stock is small. An example would be the small customer base of a mobile phone operator when there is a lack of network economies. We talk about asset interconnectedness when a lack of complementary assets impedes accumulation of an asset. The last of the above four factors, causal ambiguity, is evident when there is uncertainty regarding which specific factors or processes are required to obtain or accumulate a required asset. The question is not only which asset the firm should accumulate but also how to go about accumulating it.

Core competencies will be even more valuable when they are used as a catalyst for the accumulation of assets that are otherwise slow and costly to build. The correlation can be described thus; the more unique the customer benefits the asset in question can deliver to a market, the more valuable a firm's competence to build that asset (Verdin & Williamson, 1991).

Competence based competition is, at its core, a contest for the acquisition of skills and the development of competencies- a contest which manifests itself externally as a competition in product markets (Rumelt, 1994, p. xvi).

Researchers and managers trying to apply the concept of competencies to concrete practice(s) on the firm level are faced with a multitude of methodologies from which to choose as there is no single commonly-approved approach. However, all methodologies are drawn from the architecture of core competencies as previously presented in Figure 2. A breakdown of firm specific capabilities is thus required.

At a firm level, Hafeez et al. (2002) propose a stepwise methodology leading to the identification of a firm' score competencies. The three stages are:

- identification of key capabilities,
- determination of competence, and
- determination of core competence.

The identification of key capabilities starts by internally benchmarking key business functions of interest. Looking at a firm as a whole, these can range from general management, financial management, marketing, selling and market research, to product R&D, engineering, production, distribution and others. Any analysis should include both financial and non-financial measures. In quantitative terms, key capabilities are recognized as those that help generate high profit margins and are clear market winners in terms of securing market share. Key capabilities that are still developing, for instance through a firm's R&D, and do not yet contribute to financial results can be overlooked if only financial measures are applied. However, in this way potentially valuable dynamic competencies may be neglected. A balanced scorecard is a useful tool for the adequate capture of both financial and non-financial measures. At the next stage, the collectiveness of capabilities is assessed: that is, their integration in the company-wide business activities. The selected key integrated capabilities are further evaluated with respect to their uniqueness, i.e. their rareness, inimitability and non-substitutability. Once competencies are identified and obtained they are further analyzed for their strategic flexibility, the result of all of which is the specification of core competencies. Strategic flexibility denotes how rapidly a competence can be redeployed or reorganized for the future development of the business.

In what follows are presented two methodologies for analyzing firm-specific capabilities that underlie the core competencies of the firms in question and assessing which capabilities should be enhanced since they are vital to the pursuit of the strategy.

Chiesa et al. (1999) developed a four-step methodology for evaluating the relevance of firm-specific technological capabilities for competitive advantage achieved through R&D. The authors use term technological competencies, however, with respect to the definition used in the thesis, they are referring to capabilities. The aim is to provide an answer to the question of how to go about selecting the set of capabilities in which to invest the firm's resources. The focus is on technological capabilities alone. The methodology was applied to the company Philips, manufacturer of consumer electronic products, and involved a four-step process that begins with the mapping of the technological competencies. This stage includes the mapping of future products and the embedded technological capabilities in each scenario. Also estimated is the value of future products in terms of potential turnover, value added or margin. The final step is to assess the importance of the technological capabilities in determining the product value. The stage that follows is evaluation of the relevance of the technological capabilities. The relative contribution of each product to the total turnover is estimated. Furthermore, each capability's relative contribution to the value of the product is evaluated, and weight is assigned according to their assessed relevance. The third stage deals with the evaluation of the probability of success of the technological capabilities. Therefore, commercial risk is evaluated as the dispersion of a technological capability within different scenarios. Technological risk, on the other hand, is estimated as a function of the resource adequacy, the level of progress of the technology and the difficulty of the objectives. Finally, the success probability is evaluated. The last stage deals with the selection of the core technological capabilities. Initially, the relevance and success probability of each technological capability are jointly considered, leading to the construction of a relevance/success probability matrix and the definition of the available budget, with R&D investment taken into account. Further considerations are the firm's attitude towards risk as well as possible interdependencies between competencies. Core technological capabilities are then identified as the best-performing capabilities with high relevance and high success probability within the constructed matrix. The process concludes with the carrying out of an overall portfolio analysis of core technological capabilities is conducted with the aim of checking whether it fits with the firm's strategy.

A comprehensive methodology for identifying technological and market capabilities and their complementarities related to the R&D function at a firm level is also presented in Prašnikar et al. (2008) and applied to the case of Gorenje, a Slovenian producer of household appliances. The methodology starts with the identification of all significant technological and marketing capabilities, first at the individual strategic unit level and then at the firm level. A group of experts from the firm take part in this process. These capabilities are further evaluated, both internally and externally, in terms of the following dimensions: capability relevance, probability of technological success (for technological capabilities) or probability of attaining customer loyalty (for marketing capabilities), and competitive position. Internal analysis examines a capability's relative importance while external analysis examines the capability's competitive position relative to both the leading competitor within the industry and the general industry trends. Lastly, the methodology examines the interrelationships between the two sets of capabilities. The result of this analysis is the identification of key core marketing and technological capabilities that must be simultaneously developed and fostered within the company's overall marketing and technological strategy.

5.2 Competencies as drivers of innovation in the R&D function

Innovations along the firm's value chain are firm specific as they are based on the firm's unique way of combining resources and capabilities (Porter, 1998a, p. 124-126). Only those key capabilities that are relatively unique and common to various business functions, products and business units are likely to form and constitute the competencies of a company (Sanchez, 2004). These are industry-specific and can be identified via the use of internal and external knowledge of relevant experts (managers) (Hafeez et al, 2007).

Based on primary sources of innovation, Pavitt (1990) noted the following five distinct categories of industry: science-based (e.g. pharmaceutical), supplier-dominated (e.g. agriculture), specialized suppliers (e.g. machinery), scale-intensive (e.g. automotive), and information intensive (e.g. finance). Hay and Morris (1991, p. 243-256) further showed that within any given sector there is significant variance in the innovative performance of firms. This finding implies that firm-specific competencies are as important as technological and commercial opportunity. Tidd et al. (1996) identified significant differences in the technological and commercial opportunities of different sectors as well as in the innovative efficiency of firms within the same sector. Technological opportunity was measured as R&D spending, commercial opportunity as new products introduced and innovative efficiency as R&D spending per new product.

In general, the three broad types of core competencies are: market-access competencies, integrity-related competencies, and functionality-related competencies. Market-access competencies refer to the management of brand development, sales and marketing, distribution and logistics, and technical support, which are all those skills that help to ensure a firm's close relationship with its customers. The second type encompasses competencies such as quality, cycle time management, just-in-time inventory management and other competencies that enhance a company's flexibility and reliability relative to their competitors. Functionality-related competencies are skills which enable the company to invest its services or products with unique functionality; that is, skills which contribute to radical innovation/improvements (Hamel, 1994, p. 16).

Firms' new product portfolios strike a balance between new products based on incremental innovation and fundamental innovation (Schewe, 1996; Ali et al., 1993). The developments of new-generation products (based on radical innovations) and of products shaping new industry trends make use of substantially different and novel technologies. In the case of incremental modifications of products, "market pull" provides information on customers' preferences, while "technology push" prevails with completely new technologies that serve to address customers' latent needs (Tidd & Bodley, 2002). Since consumers buy products based on the benefits said products confer, it is still necessary for "technology push" to observe customer needs. Therefore, customer and market analysis are also crucial for technologically more novel innovations (Bacon et al., 1994).

One stream of research has identified that a combination of technological and marketing competencies creates competitive advantage (Hafeez et al., 2002; Sanchez et al., 1996; Hammel & Heene, 1994; Prahalad & Hamel, 1990). A firm with strong technological competencies is capable of using scientific knowledge promptly to develop products and processes that offer new benefits and create value for customers (McEvily et al., 2004). On the other hand, a firm with strong marketing competencies is able to use its deep understanding of customer needs to foster development of new products and organize marketing activities that provide a unique value to consumers (Vorhies, 1998; Day, 1994). In addition to each of the direct effects discussed above, technological and marketing capabilities also operate in an integrated manner (Song et al., 2005; Wang et al., 2004; Rothaermel, 2001; Fisher & Maltz, 1997). For a firm to be able to exploit its competencies fully through innovation, investments must be made in complementary "assets" or knowledge of tools, methodologies and process that can facilitate this (Tidd, 2006, p. 12).

The knowledge represented by these competencies contributes to the speed and flexibility of the development process and results in competitive products. As suggested by Swink and Song (2007), both marketing and technological capabilities have a substantial impact at each stage of new-product development, which is in turn associated with higher project return on investment. Competencies not only influence product's competitive advantage but also project lead times. The manner in which specific groups of competencies contribute to different stages of new product development is summarised in Table 1 and addressed in detail in separate chapters on the distinct competencies in question.

Table 1: Competencies employed at different stages of new product development

		New product development stages			
		Busness/market analysis	Technical development stage	Product testing	Product commercialization
Competencies	Technological competencies	Technical feasibility of products	Engineering studies, establishing product designs, prototyping	Influencing consumer tests design and results interpretation	Production plans and ramp-up
	Marketing competencies	Evaluation of market impacts of product feature options	Facilitating product feature decisions	Sample selection, testing, analysis	Marketing plans, product promotion, distribution
	Complementary competencies	Aligning new product features with potential customers' needs, assessment of needed investment and risks	Alignment of technological and marketing knowledge	Translating testing results in design modifications	Coordination of production planning and demand management activities

Sources: Adapted after Swink & Song (2007), Coates & McDermott (2002), Fowler et al. (2000).

Further support for the concept of competencies and their contribution to competitive advantage can also be recognized in the ideas put forward by Amar Bhidé in his book 'The Venturesome Economy' (2008, p. 272-286). Therein he posits that inventions and ideas can easily travel across national borders while commercialization, diffusion and use of inventions are of more value to companies and societies. He attributes the decisive advantage of the USA over its rivals- including Japan, which began catching up in terms of technology in the 1980's- to sophisticated marketing, distribution, sales and customer-service systems. In fact, this idea is also quite closely related to the nature of marketing and complementary competencies, taking into account the importance of market insight alongside technological superiority.

Studies also imply a significant link between product quality and product innovation. From a theoretical point of view, any kind of improvement in product quality is, to a certain degree, reflected in the development of new products and can be considered an innovation, for example, a change of materials used or a change in the technological or mechanical design of the product. Kano et al. (1984) claim this is especially true when the elements of the product quality focus on the 'delighting' level beyond the basic and stated levels of customer needs and expectations. As far as product innovation based on exploiting new technologies is concerned, several aspects of product quality tend to be improved. Prajogo et al. (2008, p. 629) emphasize that improved quality of the product must be inherent (i.e. assumed) in innovation.

5.3 Previous empirical studies

There is a vast body of research on competencies and underlying capabilities. In this chapter is presented an overview of some of the most representative studies, which are also summarized in Appendix B.

Hitt and Ireland (1985), showed by means of a sample of 185 Fortune 1000 industrial firms that there is a link between corporate distinctive competencies and firm performance. On the basis of a literature review they compiled a working set of 55 distinctive competencies. Common to all 55 was the fact that they occur through the development of specific activities associated with 7 business functions, namely: general administration, production/operation, engineering and R&D, marketing, finance, personnel, and public and governmental relations. They go on to posit that firms must develop synergies among their business units and should not be viewed as a portfolio of unrelated business units. One way of developing synergy is through the transfer of corporate-wide distinctive competencies between the units. Distinctive competencies facilitate the implementation of a firm's grand strategy, whether the focus is on stability (similar operating levels through incremental performance improvements), internal growth through internal development, external acquisitive growth (through acquisition, merger or joint venture) or retrenchment, which refers to a reduction of the scope-level of product/market objectives. Respondents were firms' CEOs and senior executives, the former denoting the firm's grand strategy while the latter, who are knowledgeable about overall firm operations, provided the rest of the answers. Each of the 55 activities was ranked on a seven-point scale according to their strategic significance. The relative importance of each group of distinctive competence activities was obtained by aggregating individual results. Firm performance was measured by market returns. Results show that firms pursuing a stability strategy focus on marketing by improving distribution networks and by developing effective policies for product additions and deletions in order to achieve sales levels that best make use of plant capacity. An internal growth strategy is mostly dependent on financial control of operations and negatively correlated to engineering and R&D. This negative relationship is connected to poor R&D management. A strategy of acquisition is linked to production/operations activities and retrenchment strategy to reductions in the objectives and/or scales of a given operation. The authors also distinguished between 4 industries, namely consumer non-durable goods, consumer durable goods, capital goods, and producer goods. Engineering and R&D were negatively related to markets for consumer non-durable goods, which tend to be highly competitive more often relying on competing based on price than quality. No relationships were established for consumer durables. Capital goods, on the other hand, are often custom manufactured at fixed contract prices. Thus, establishing and maintaining firm efficiency by controlling production costs and meeting customer requirements is crucial. Producer goods are sold on a business-to-business basis to be integrated into final products. Manufacturing efficiency and quality control have greater importance than marketing activities, such as differential pricing strategies and advertising.

In his 1996 study, Chang (1996) investigated the impact of technology and marketing competencies on profitability and firm performance. Chang uses the term capability for the concept defined in the present thesis as competency. Using data from the PIMS database for 2744 firms from the USA, Canada, the UK and EU, he showed that technology and market competencies contribute significantly to a firm's ROI, ROS, cash flow on investment and market share. As indentified by the database, 52% of the firms in the sample were identified as market pioneers, with the remainder (48%) being market followers and late entrants. His analysis revealed that market pioneers possess significantly higher technology and marketing capabilities than market followers. The study of competencies is, however, limited by the number of indicators available from the PIMS database and the different scales in use. Technology competencies were thus measured using product change frequency and new product development time as proxies of product improvement and product quality as a proxy of manufacturing competence. Measures of marketing competencies included assessment of product breadth, percentage sales from new products, price, sales force expenses, advertising expenses, promotion expenses, services, image and forward integration used to represent or denote distribution channels. Specific measures were calculated as averages from the data over the period of the previous four years. Technology and marketing competencies were finally calculated by standardizing the sum of the measures and subsequently used in OLS regression. Synergies between technology and marketing competencies were also investigated. An interaction term was included in the model. Although there is no synergy effect on ROI, ROS and the ratio between cash flow and investment, there is a positive effect on a firm's market share. Synergies appear to help the firm cope better with market conditions.

Based on a synthesis of the existing literature, Fowler et al. (2000) propose that market-driven, technological and integration competencies are central to the creation of competitive advantage in dynamic environments. They suggest that in such environments new opportunities should be exploited through the above mentioned three groups of competencies instead of product-centred strategies. Technological competence is the "ability of the firm to combine knowledge about the physical world in unique ways, transforming this knowledge into designs and instructions for creating desired outcomes." Customer knowledge, customer access and competitor knowledge are referred to as the three main elements of market-driven competencies. Following Grant's definition (1996), integration competencies enable the firm to combine the wide-ranging capabilities, information, and perspectives necessary to develop successful products. Development of technological and marketing competencies is very much influenced by a firm's absorptive capacity, which increases its ability to recognize and apply new external knowledge in order to continue the firm's competence development. The authors do not develop a competence-measurement model but provide a list of potential measures of the proposed constructs. As possible measures of market-driven competencies, the following are offered: spending per customer, number and percent of repeat customers, referred customers, customer complaints, response to customer requests, punctual delivery, number of competitors serving the same customer and a profile of competitors' market competencies. For the purpose of measuring technological competencies they suggest: cycle time, unit cost, yield, set-up time, common parts/common technologies, number of competitors able to produce the same specific technology, and profile of competitors' technological competencies. To capture integration competencies they put forward as potential measures: product profitability, percent of sales from new products, variety of products, warranty costs, cost of quality as percent of sales, actual introduction schedule versus plan, number of competitors delivering similar products, and a profile of competitors' integration competencies.

An exploratory, within-case, longitudinal study of an emerging technology project undertaken by the large US high-tech manufacturing company Coates and McDermott (2002) reveals that technology, market and integration competencies are the three groups of competencies that were newly created in support of the development of emerging technology. The development process took the company into areas in which it had limited knowledge concerning new technology and its potential applications. These new competencies helped the firm develop attractive product market positions and gain the advantages of a first-mover. Data were obtained through both structured and unstructured interviews with the managers, engineers and scientists most actively involved in the development. Comparisons of the responses led to the identification of critical capabilities comprising new competencies. The structured elements included four questions, the first of which addressed the capabilities necessary to develop and compete in the new technology market. Further listed items were knowledge and skills used, together with the development outcome(s) they facilitated. As competencies spread across business units, respondents were asked to identify capabilities that would still be used even if the new division were removed. Lastly, capabilities were compared to those of other firms. Technology competencies stemming from the understanding of the new technology involved design and manufacturing skills, equipment, know-how or processes. These competencies enable manufacturing flexibility and contribute to the reliability of products and their manufacturing processes. Market competencies include managing the perceptions of current and potential customers, choosing the right customers and, subsequently, building relationships with them. Integration competencies were identified as those positively influencing problem solving and the combining of different knowledge areas.

On a stratified sample of 248 high-tech firms in China, Wang et al. (2004) demonstrate that marketing, technological and integrative competencies have a significant influence on firm performance. They define technological competencies as those that determine which products or services can be provided technically at one time. Marketing competencies determine which products or services demanded by targeted customers can be detected. Integrative competencies reflect the degree of fit between technological and marketing competencies, as well as the efficiency with which products of customer value are delivered. They argue that although much of the research on a firm's core competencies emphasizes the role of technological and marketing competencies, it is the integrative competencies that enable the firm to deploy its unique resources and capabilities in such a way as to respond successfully to various changing environmental conditions, thus achieving sustainable performance. Measures were developed based on field research and expert group consultations. Chief executive officer or company presidents took part in the survey based on a structured questionnaire. A seven-point Likert-type scale was used with a ranking system ranging from "absolutely disagree" to "totally agree". Although all firms included in the sample were high-tech firms, they came from different industries: computer related products, electronics, electric equipment, telecommunications equipment, and pharmaceuticals. Marketing competencies were based on the measures of the following eight capabilities: access to information on customers, communication with customers, customer involvement, responsiveness to customers, information on competitors, benchmarking of products and services, marketing channels, and managing of customer relationships. Eight technological competencies encompassed R&D investments, technological skills, attracting and motivating experts, the prediction of technological trends, the application of new-technology in problem solving and industry leadership. The integrative competencies measured were as follows: the ability to communicate among and between functions, leveraging of marketing and technology knowledge, the integration of external and in-house resources, leveraging of competitors' strategies, the use of new technological findings, the integration of customers' innovative ideas, the delivery superior value by process integration, and coordination in the implementation of the corporate strategy. Firm performance was measured by respondents estimating how satisfied they were relative to major competitors in terms of return on investment, market share, customer value and cost effectiveness. The authors validated the model using the Partial Least Squares approach to structural equation modelling. Furthermore, technological and market turbulence proved to be a strong moderator of the relationships between the competencies and firm performance, though market turbulence had no observable effect on the relationship between integrative competencies and firm performance.

Studying technological and network competencies, Ritter and Gemünden (2004) found that both technological and network competence affect innovation success and contribute to strategic flexibility. Network competence enables a firm to establish and make use of relationships with other organizations. The authors view this competence as an extension of marketing competencies, arguing that it highlights the interaction by which firms acquire information, exchange offerings and collaborate technologically. Network competence was measured in terms of the intensity of networking in business activities and by the extent to which employees participating in these networks possess special and social qualifications. Their definition of technological competence refers only to a firm's internal understanding and the exploitation of the relevant state-of-the-art technology. It encapsulates four grounds for technological collaboration and four statements regarding technological expertise. Innovation success was divided into three product innovation measures and three process innovation measures. Seven-point Likert-type scales were used. The model was tested using structural equation modelling and LISREL software. Business strategies which were analyzed only in the context of technology (defined as the importance of R&D and new-product development and the desire to be the technological leader in the market) were not directly related to innovation success but support development of both groups of competencies. The sample was comprised of 308 German firms from the industries of mechanical and electrical engineering. Industry-specific or environmental characteristics were not included in the model.

Lokshin et al. (2009) devised a structured questionnaire for the purpose of measuring customer, technological and organizational competencies and their respective impacts on innovative performance. Customer competence was measured by market research, customer cooperation and customer sourcing. The indicators of technological competence employed were monitoring, transfer, quality control and intellectual property. Organizational competencies refer to organizational practices that have been identified by previous research as fostering firm innovativeness. They were measured with reference to two indicators; team structure (the ability to build and maintain team cohesiveness) and slack time (as a way of promoting business creativity by giving the employees a certain amount of autonomy). Likert scales were again used to evaluate competencies. Innovative performance was measured by the number of successful product innovations realized by a firm in the previous two years and whether the firm had realized radical innovations during the same period. Data was gathered for 27 German firms operating within the fast-moving consumer goods industry. The authors confirmed the direct effect of organizational competencies on innovative performance through the use of regression models. The synergetic effect of combining technological, customer and organizational competencies on product innovation was also demonstrated. This effect is especially significant for radical innovation. Moreover, higher levels of competencies are characteristic of firms with higher innovation output. Radical innovations also require higher levels of firm competencies than is the case for incremental innovations.

5.4 Technological competencies

Technological or technology competencies incorporate practical and theoretical know-how, as well as the methods, experience, and equipment necessary for developing new products (Wang et al., 2004). They encompass a portfolio of technological capabilities concerning the capacity of the company to utilize scientific and technical knowledge for the research and development of products and processes, which, in turn, leads to enhanced innovativeness and performance (McEvily et al., 2004). According to Swink and Song (2007) technological competencies influence all four stages of the new-product development process. At the first stage of business/market analysis technological competencies help address the technical feasibility of the products in question. The technical development stage incorporates product- and process-engineering studies and continues with the establishing of product designs and specifications, the prototyping of the product and the approving of final designs. In all of these tasks technological competencies have a central position. During the third stage of product testing technological competencies are of secondary importance. Nonetheless, they continue to influence the design of consumer tests and the interpretation of results. At the final stage of product commercialization they are key elements, both for production plans and production ramp-up.

To reflect the construct of technological competencies various qualitative indicators are in use as there is still no commonly accepted methodology. Studies rely on self-assessment scales either by stating agreement with performance statements (Lokshin et al., 2009; Belderbos et al., 2004; Fritsch & Lukas, 2001) or comparative evaluations relative to competitors (Wang et al., 2004; Danneels, 2002; Torkkeli and Tuominen, 2002; Afuah, 2002; Walsh and Linton, 2002; Tyler, 2001; Kumiko, 1994). Measures of technological competencies are incorporated in statements and cover the following:

- investments in R&D activities,
- the accumulation of stronger and more diverse technological skills,
- the provision of on-the-job training to improve the technical skills of employees,
- attracting and motivating talented experts,
- the ability to predict future technological trends accurately,
- skills in applying new technology to problem-solving,
- industry leadership in establishing and upgrading technology standards,
- technological leadership in the principal industry,
- the monitoring of product areas outside the company (e.g. what other companies in the same industry are doing; what consultancy firms are currently recommending) to find out whether the technology is up to date, and
- the monitoring of the employees involved and the process' outcome.

The number of new patents, copyrights, registered trademarks, or registered designs that have been successfully applied for within a period is still often included as an indicator of technological competencies although they are generally considered to be innovation outputs and, therefore, measures of innovative performance.

Ivarsson and Jonsson (2003) analyzed the technological competence of transnational companies in asset-seeking direct foreign investment. Using unique firm level data pertaining to 231 majority-owned foreign affiliates located in West Sweden in the manufacturing and wholesale industry, they showed that technological competencies act as an important pull-factor for asset seeking direct foreign investment in a small developed economy.

5.5 Marketing competencies

The role of marketing along a firm's values chain is critical, especially from the viewpoint of the relationship between a firm and its customers in the pre-development and post-delivery stages. Within the Total Quality Management business management strategy, some authors emphasize customer focus as being the starting point of the quality philosophy (Deming, 2000; Juran, 1989; Crosby, 1995, p. 192). Marketing is namely expected to close the so called quality gap between what customers want and what they receive. It also enters the process at the very beginning and is the initial point of contact with the customer.

There is a stream of studies within the literature which argues that the understanding of market needs is of paramount importance to innovation success (Slater & Narver, 1994; Schewe, 1994; Flores, 1993). Furthermore, the relationship between customer orientation and organizational innovation has also been confirmed (Lukas & Ferrell, 2000; Appiah-Adu & Singh, 1998).

A special role is also played by suppliers as they perform activities, and incur costs, when creating and delivering the purchased inputs which are used in a firm's end product(s). Their involvement can range from simple consultation concerning design ideas to full responsibility for the design of components or systems they, as suppliers, will provide. The incentive for closer supplier collaboration is provided by the possibility of helping suppliers reduce their costs or improve the quality and performance of the supplied materials, all of which improves a firm's competitiveness, contributing to a firm's cost- and product-differentiation capabilities (Prajogo et al., 2008, p. 621). Deming (2000) makes a case for the theory that certain US firms make decisions regarding purchasing and supplier selection based solely on price, which inevitably results in the frequent changing of suppliers. What firms should be aiming for is building cooperative relationships with suppliers by developing joint quality improvement programs and, therefore, entering long-term contracts with those suppliers in order to allow them to make greater commitment to improving the input-product quality. In return, firms can reduce their supplier base and save on administrative costs as well as improve quality variability. Also recognized was the significant contribution of suppliers with regard to innovation performance. Handfield et al. (1999) observe in their study of supplier relations that although 45% of the firms in their sample were not satisfied with their current supplier relations, they did recognize this factor as being of continuing importance and consequently planned to commit to further supplier-integration. A critical factor in success is how well the firm understands supplier's capabilities, ranging from the supplier's ability to meet cost, quality and ramp-up goals, and how well they are able to assess the technology roadmap, to their level of design expertise and the volatility of change in the particular technology. Bozdogan et al. (1998) posit that firms should pro-actively integrate suppliers at an early stage in the concept exploration and definition stages of product development.

Companies with highly developed marketing competencies are well aware of customer needs and are capable of value creation with respect to all elements of a product or service that are relevant to the customers (Day, 1994). Constituent marketing capabilities are therefore an interwoven system based on knowledge and skills that allow the company to generate customer value and also facilitate timely and effective responses to marketing challenges (Song et al., 2005; Vorhies & Harker, 2000; Vorhies, 1998). At the business/market analysis stage marketing competencies provide an evaluation of the market impacts of product-feature options (Kahurana & Rosenthal, 1997) as the aim is to understand the competitive positioning of the future product. During the technical development stage marketing competencies facilitate product feature decisions. Marketing usually takes a leading role in product testing, which encompasses the selection of key customers and sites, testing of markets and result analysis. Marketing plans, product promotion and distribution are tasks that require marketing competencies for successful product launches at the product commercialization stage (Swink & Song, 2007; Paul & Peter, 1994).

Examples of measures of marketing competencies to be evaluated on scales expressing the extent of agreement (Lokshin et al., 2009; Thomke & von Hippel, 2002; Tether, 2002):

- cooperation with customers regarding product innovation occurs on a regular basis,
- reliance on market research when developing a new product or product feature,
- customers as a source of ideas for new products, and
- acquainting oneself with customers and their needs to find out what products they will need in the future.

Statements about marketing competencies to be assessed relative to competitors refer to (Song et al., 2005; Wang et al., 2004; Li & Cavusgil, 2000; Vorhies et al., 1999; Li & Calantone, 1998; Tuominen et al., 1997; Day, 1994):

- obtaining real-time information about changes in customer needs,
- communicating with customers about their potential and current demands,
- the involvement of customers in the process of product testing and assessment,
- the degree of responsiveness to customers' requirements,

- the acquisition of real-time information concerning competitors' evolution of strengths and weaknesses,
- benchmarking of the product and service practices of major competitors,
- building and enhancing marketing channels, and
- creating and managing close/durable customer relationship effectively over the long-term.

5.6 Complementary competencies

Some authors treat complementary capabilities and competencies as an interaction between technological and marketing capabilities and competencies. For example, Song et al. (2005) showed that the interaction effect on business performance was significant within a high-turbulence environment only. However, complementary competencies are now gaining increasing recognition as an independent group. They reflect the degree of fit between the two groups and should be treated as a distinct network of capabilities. A failure to value them properly can lead to the inadequate identification of key capabilities. In the literature they are also referred to as integrative, integration or combinative competencies.

The role of complementary competencies, according to Wang et al. (2004) is to:

- integrate different technological specialties,
- combine different functional specialties,
- exploit synergies across business units,
- combine in-house resources with the external capabilities required, and
- integrate the dynamic competence building process to bring about superior performance.

The alignment of new product features (technological aspect) with potential customer needs (marketing aspect) is the role of complementary competencies at the first stage of new-product development. They are also employed in the assessment of the investment required and the evaluation of accompanying risks (Swink & Song, 2007). Similar complementarity of technological and marketing knowledge is also crucial during the second stage of technical development. At the same time, it has proven to be positively related to the translation of testing results into product and process design modifications (Song et al., 1998) during the product testing stage. The integration of both streams of competencies contributes to an improved coordination of production planning and demand-management activities during product commercialization.

Firm practices in new product development also point to the importance of joining technological and market knowledge, a process which leads to higher product quality. These elements lie at the centre of the quality function deployment practice- the origins of which can be traced back to late 1960's Japan- a practice in which consumer needs and competitive evaluation present a basis for the identification of the technological requirements of a product (Akao, 2004, p. 3-7). Similarly, concurrent engineering promotes the effective coordination of the activities of different departments and encourages cross-functional teams throughout the process of new-product development (Prasad, 1996, p. 198). It was high-technology firms that first actively looked to advance cross-functional management processes, focusing primarily on integrating product development, product strategy and the supply-chain (Goffin & New, 2001). A conspicuous trend which provides broader support for this line of thinking can be observed in the fact that large, well-established technology firms are relying less and less on traditional, big R&D laboratories and are placing an increasing amount of emphasis on development and the ability to respond quickly to needs emerging on the market (The Economist, 2007).

The competence of combining in-house and external resources or taking part in strategic technological alliances draws from the aspect of competence-based competition that regards firms as open systems (Sanchez & Heene, 1997). Through linking resources within networks, cooperating firms may jointly realize the benefits of asset-mass efficiencies, asset interconnectedness and reduced time compression "diseconomies" that would otherwise be unavailable to the firms as stand-alone organizations (Dierickx & Cool, 1989). A study by Gupta and Wilemon (1996) based on the experience and ideas of 120 R&D directors showed that both vertical and, increasingly, horizontal collaborations in R&D activities can lead to a more efficient R&D function and, consequently, better business results. The prerequisite is, however, a close link between technology and strategy, which points to R&D being more business- than technology-driven. Chesbrough (2003) coined the term open innovation to describe opening up in-house R&D to the external environment. He states that in the past internal R&D was a valuable strategic asset, often acting as a barrier to competitors entering many markets. The only firms that were able to compete in terms of R&D within their industries were large corporations. Potential competitors had to make heavy initial investments in their R&D facilities in order even to be in a position to try to compete. Contemporary patterns of competition show that these once-leading industrial enterprises are now encountering very strong competition from many start-ups. However, these newcomers conduct little or no basic research of their own, but instead carry out R&D via strategic technological partnerships. Possible forms of cooperation include either the more popular contractual partnerships or equity-based joint ventures (Hagedoorn, 2002). Vertical partnerships involve the cooperation of partners from along the value chain. A firm can thus collaborate with either suppliers or customers. Horizontal partnerships are collaborative R&D projects carried out by close or more distant competitors. A special and noteworthy kind of partnership is that set of collaborations which involve cooperation with public research institutions and universities (Backes-Gellner et al., 2005).

While the external environment can refer merely to the outside environment of the firm in question on a national level, findings show that multinational firms that took a global approach to research outperformed those that concentrated their research activities only on their domestic market (Jaruzelski & Dehoff, 2008).

Examples from practice further demonstrate that the establishing and maintaining of a competitive position derives from complementary capabilities as building blocks of complementary competencies (Rothaermel, 2001; Eisenhardt & Martin, 2000; Grant, 1996; Kogut & Zander, 1992). As a result of their usually tacit nature, they are difficult to identify, observe and articulate. Numerous studies confirm that complementary capabilities facilitate synergies between technological and marketing capabilities, consequently generating new applications of the existing knowledge (Song et al., 2005; Peteraf, 1993; Barney, 1991). Even though companies might have unique core technological and marketing capabilities or systematically develop the portfolio of their capabilities, this does not automatically translate into them outperforming their competitors.

The following are examples of measures in use that cover complementary competencies (Wang et al., 2004; Kogut & Zander, 1992; Dosch et al., 1999):

- communication among functions in the process of product and service design,
- sharing and leveraging marketing and technology knowledge among functions/business units,
- the integration of external resources with the in-house resources,
- sharing and leveraging information about competing strategies of major competitors,
- the coordination and integration of activities of functions/business units within corporate strategy,
- embedding newly achieved technological findings in new products and services,
- incorporating customers' innovative ideas in final products and services,
- delivering superior value to customers via the integration of different processes, and
- the effective coordination of corporate strategy in the implementation process.

6 Model of competencies as antecedents of innovative performance and subsequent effect on business performance

6.1 Operational model

On the basis of the conceptual framework pertaining to the influence of technological, marketing and complementary competencies on innovative performance and business performance, the following operational model can be constructed (Figure 4). The model draws from the theoretical background presented in the first part of the thesis.

The focus of the empirical research will be the three groups of competencies that were established through the review of relevant existing theory; more specifically, the theories of endogenous growth and innovation as well as that of competence-based competitive advantage and innovation management being the key firm leverages in new-product development and R&D activity as such. It is of interest to further investigate technological, marketing and complementary competencies as drivers of innovative performance.

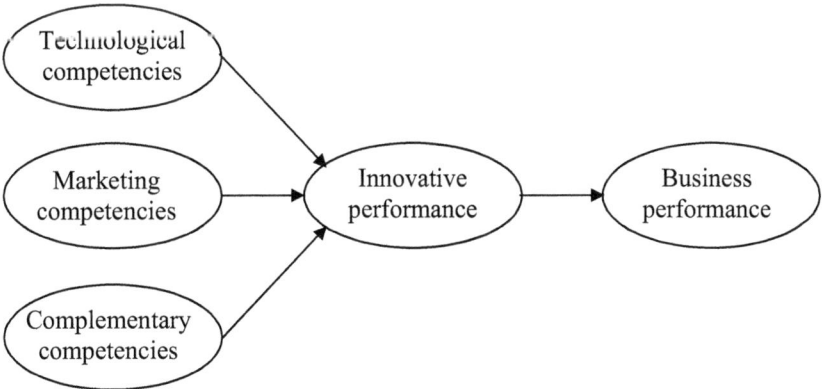

Figure 4: Operational model of the influence of technological, marketing and complementary competencies on innovative performance and business performance

More precisely, the purpose is to study the relationship between competencies and innovative performance, as existing studies focus rather on the relationship between competencies and business performance alone or else assess competencies in projects of new-product development. A small number of empirical studies (Hagedoorn & Cloodt, 2007; Song et al., 2005; Wang et al., 2004) tried to identify the various sources of superior firm performance through distinguishing different elements of core competencies and thus provided an insight into the underlying determinants of innovation and, consequently, innovative performance. Moreover, a few empirical studies can be found that examine the major constituents of core competencies and their differentiated influences on overall firm performance (Wang et al., 2004). Such research is needed in order to achieve an in-depth understanding of how and why core competencies contribute to firm performance in contingent contexts. Moreover, a study can provide insights into how firms can adapt quickly and effectively to the increasingly changing nature of both internal and external business environments, without focusing solely on the technological aspect of innovation activities. The objective of the thesis is to develop and test the model of relationships between competencies and innovative performance by controlling for industry specifics.

The main hypothesis herein is:

Hypothesis 1: *Innovative performance is affected by three groups of competencies – technological, marketing and complementary.*

According to the EIS study, Slovenia falls into the group of modest imitators with regard to innovative activity (Eurostat, 2007). Imitation is recognized as a strategy of technology following firms that requires comparatively little technological knowledge but strong competencies with respect to competitor analysis and intelligence, cost cutting, and learning in manufacturing (Porter, 1998b, 171-172). On this premise- regarding the differences between the competencies being developed by technology leaders and followers - stands the first partial hypothesis.

Hypothesis 2: *Technology-following firms have, compared to technology leaders, relatively more developed marketing and complementary competencies than technological competencies.*

Since competence building requires strategic commitment, not all companies can be expected to possess competitive competencies with respect to their competitors.

Hypothesis 3: *Among technology followers there are followers with competitive competencies and those with obsolete competencies.*

The difference between technology leading and following firms is reflected also in their innovation strategy and new product development.

Hypothesis 4: *New product development activities of technology followers rely on incremental innovation and imitation.*

New products, whether of an incremental or radical nature, are a way for firms to differentiate themselves from their competitors. Firms can decide to pursue different innovation strategies which are dependent on their competencies.

Hypothesis 5: *Radical innovations require stronger technological competencies than incremental innovations.*

Hypothesis 6: *Radical innovations are highly dependent on advanced technological knowledge.*

Hypothesis 7: *Access to external sources of knowledge is an important complement to in-house knowledge in innovation activities.*

Complementary capabilities and competencies are traditionally understood as referring to an interaction between the technological and marketing competencies (Song et al., 2005). Although they do reflect the fit between the other two groups of competencies, their role is not only to act as an intermediary between technological and marketing competencies but also to enhance them or, in other words, complement them. From this reasoning stem the following hypotheses:

Hypothesis 8: *Complementary competencies are a distinct group of competencies.*

Hypothesis 9: *Interaction of technological and marketing competencies cannot replace complementary competencies in the model of innovative performance.*

Firms aim to develop products in order to satisfy customers' needs in a novel or improved way. Increased product variety - due to introductions of new products - along with improved quality will therefore better address market needs and consequently be reflected either in higher price premiums or increased sales, thus affecting the overall business performance of a firm.

Hypothesis 10: *Innovative performance has a positive impact on the business performance of a firm.*

Hypothesis 11: *The relationship between innovative performance and business performance is moderated by environmental effects.*

Innovations can be a way for a firm to respond to technological turbulence.

Hypothesis 12: *Higher technological turbulence acts as a positive moderator of innovative performance on business performance.*

High market turbulence significantly raises uncertainty levels on the market for new innovations as they enter the market with a time lag.

Hypothesis 13: *Higher market turbulence acts as a negative moderator of innovative performance on business performance.*

6.2 Methodology and survey design

Due to the novelty and specifics of the developed model, it cannot be tested using existing datasets. Although qualitative, survey-based data is, to some extent, available for innovative performance measures, for example the Community Innovation Survey by Eurostat as presented in chapter 3.4, there is no such systematic national or cross-national survey that also incorporates questions concerning firm competencies. Therefore, one aspect of this research is also an attempt to devise a survey that could be used for this purpose. Survey design along with the resultant questionnaire is presented later on.

To test the hypotheses and operational model presented above, a set of different statistical tools is employed. The analysis starts with a descriptive analysis and the description of the sample with aggregate data for different firm characteristics. Identification of different firm segments follows. For this purpose, a segmentation based on innovative performance is carried out. Differences between segments with respect to their competencies and innovative performance are investigated. Here, support for Hypothesis 1 is already assumed. That is to say, as technological, marketing and complementary competencies are expected to have an impact on innovative performance, firms with different levels of innovative performance are expected to demonstrate different levels of competitiveness in terms of their competencies. Clustering technique with a two step methodology is applied. This technique proposes improving the segmentation initially obtained by hierarchical clustering methods via the additional application of non-hierarchical methods in order to optimize the classification of observations. The firm segments obtained are described in terms of their innovation strategy and differentiated as technology leaders or followers. Comparisons of the clusters based on significant differences between them provide evidence for Hypotheses 2, 3 and 4 regarding technology followers.

The second part of the empirical analysis is dedicated to structural models, wherein the relationships which hold between competencies, innovative performance and business performance are established. Partial Least Squares technique for structural equation modelling is used. First are tested models of innovative performance, adapting the general baseline model for incremental innovation, radical innovation and trend-setting firm strategies of innovation. The confirmation of the validity of the models provides further support for the first hypothesis. Besides establishing the links, there is also a special emphasis on the differences between the models as their implications are key for innovation management within both technology-leading and -following firms. Here is found support for Hypotheses 5, 6 and 7 regarding the differences between competencies engaged in incremental or radical innovation. Hypothesis 8 and the validity of the concept of complementary competencies are also addressed.

Testing of Hypothesis 9 using a structural model follows next. In the model complementary competencies are replaced with the interaction of technological and marketing competencies. The analysis continues with the validation of the extended model of competencies and innovative performance by including the link between the innovative performance and business performance of a firm, thereby testing Hypothesis 10. Furthermore, possible sampling bias is checked. Sampling bias may occur due to the fact that specific types of respondents are more willing to participate in the survey than others. However, should the samples be biased, this would mean the obtained models are too, thus necessitating their correction. In the end are introduced two more concepts to the model. These are two external environmental effects, namely, technological and marketing turbulence, which according to the literature can be expected to have an impact on the link between innovative performance and business performance. Hypotheses 11, 12 and 13 are tested via this model.

As previously mentioned, there is no existing national or cross-national public innovation survey that would systematically collect data on firm competencies which required a custom survey design and questionnaire. Designed was a questionnaire to best suit the operational model. The variables required to simulate the proposed theoretical concepts were selected on the basis of economic, organization and management literature. A multi-industry sample of Slovenian manufacturing firms was targeted.

There are several reasons why a multi-industry sample was chosen. The first of these lies in the definition of competitive advantage and core competence. Core competencies apply to more than one core product and consequently more than one business unit. Following this definition, core competencies are presented as a level of analysis and investment superior to the level of products and markets (Tidd, 2006, p. 6). Secondly, the multi-industry approach was chosen due to the diversification of large companies. As firms attempt to take advantage of synergies and economies of scale and scope, many diversify into different businesses. Research on diversification in production in developed countries shows both that big firms are more diversified than small firms, and that more diversified firms demonstrate greater R&D intensity than those less diversified (Gollop & Monahan, 1991; Lichtenberg & Siegel, 1991). Products becoming more and more multi-technological also require companies to develop competencies in an increasing range of technological fields in order to maintain their competitiveness (Tidd, 2006, p. 9, Markides & Williamson, 1994). At the same time, the segmentation of markets that firms within the same industry serve can be so fragmented and diverse that it is often difficult to pool companies within the same industry together based solely on formal classifications of their core business. The key reason, however, is the aim of differentiating firms that are technology followers from technology leaders within the economy. In order to control for industry characteristics, environmental specifics are also considered as moderating variables in the analysis. The selected indicators of the concepts included in the model and questionnaire thus enable a multi-industry analysis of the manufacturing sector.

The manufacturing industry alone has been chosen since their innovation activities as well as value chains are more standardized than those in the services sector and thus make inter-firm analysis with corresponding comparison easier and more straightforward. This is discussed in more detail in the chapter on innovation in services. Given the nature of the present research, more specifically the nature and scope of competencies, it can be expected that in accordance with the assimilation approach of service industry analysis, conclusions are also of relevance to firms in service sectors.

Companies included in the survey were classified according to size; medium-sized and large. These companies are more likely to have systematically organized R&D functions and are continuously forced to innovate in order to sustain or improve their competitive position and withstand dynamics in industries, unlike small firms whose innovative products often cater for small niche markets. Furthermore, Tidd et al. (1997) report that many small and medium sized firms fail to innovate on time since they seem to be caught up in the vicious circle of being fully occupied with solving short-term operational problems. Consequently, management teams pay less attention to their long-term strategy and remain stuck in operational problem solving.

Along similar lines, it has been found that within the developed world a large number of newly established companies are unable to survive the first few years of their existence (Caves, 1998; Geroski, 1995). A report on the Dutch economy reveals that as many as 40% of newly established companies were unable to survive the first five years of operation (Ministry of Economic Affairs, 1996; Geroski, 1995). Small and young firms in particular are at most risk of exit (Cefis

& Marsili, 2006). They are, at the same time, less likely to have formal R&D laboratories. Even when they carry out their R&D activities in-house they usually do not record them in their profit and loss accounts (Patel & Pavitt, 1995). In order to draw on systematic experience of firms, the additional restriction that companies had to have been active for at least the past five years was imposed.

The structured questionnaire is designed in a way that acknowledges the funnel approach (Bickart, 1993), meaning general questions are followed by progressively more specific questions. Such sequencing prevents specific questions from biasing responses to the general ones. An English translation of the questionnaire is included in Appendix C.

First, participating firms are asked to list their product lines (Question Q A.1.). A literature review of cross-industry studies has shown that none have thus far carried out analyses of either companies' capabilities or competencies for specific product lines. However, the intention is to account in the thesis for the product diversification that has also proven to be an important facilitating factor during the new-product development stage. Firms may enter new lines of business through either internal business development or acquisition (Ramanujam & Varadarajan, 1989). The main reasons for doing so can range from perceived benefits associated with a greater target market and the utilization of unused productive capacity, to risk reduction from the viewpoint of diverse business portfolio and capability build up (Chakrabarti et al., 2007; Montgomery, 1994). Consequently, not all product lines within a company draw from the same set of capabilities. Potential synergies between diverse product lines often encourage firms to opt for diversification. Nevertheless, resources, knowledge and technologies belonging to a specific product line can be very specific. In addition to this, the diverse products may be present in very diverse and specific markets. This makes any generalizations made across product lines less reliable. If companies estimated that their product lines could not be analyzed together (Q B.1.), they listed them separately and provided separate answers for every product line or group of alike product lines. Similarly, the Strategic Planning Institute (The Strategic Planning Institute, 2008) uses strategic business unit as the unit of analysis for its PIMS (Profit Impact of Market Strategies) database. Each business is a division, product line, or other profit centre within its parent company. Firms are also asked to provide the tenure of the company in its core industry. It is important to keep in mind that many companies were restructured as new legal entities after Slovenia gained its independence. Therefore, many of the companies looked at were officially founded in the early 1990's, although they may have a much longer tradition in the industry.

The first set of questions (Q B.2.) is dedicated to industry characteristics; more specifically, indicators of market and technological turbulence. Four different indicators were applied to each category of environmental turbulence (Wang et al., 2004; Calantone et al., 2003; Song et al., 2005). In the case of technological turbulence, the elements measured were: the speed of change in technology; opportunities arising due to new technologies; the ability to predict technological change, and the extent of technological change in the industry. Questions regarding market turbulence referred to: market uncertainty; the predictability of changes in demand; the predictability of competitors' activities, and competition intensity. Answers are ranked on a five-point Likert scale.

Section Q B.3. follows with statements regarding competencies and innovative performance. This set of variables is more closely discussed in the following chapter on variables.

Quantitative data on innovative performance was captured also by: R&D expenditure, patent counts, model counts, recently obtained patents, share of new products in total sales, and awards for products (Q C.4., C.5., C.7., C.9., C.10.). Due to the previously discussed shortcomings of quantitative measures of innovative performance, the model was built on qualitative measures, as further explained in the following chapter.

Literature on new product development deals with different aspects of R&D function (Griffin, 1997; Cooper & Kleinschmidt, 1995). Included in the questionnaire were questions on strategic cooperation in R&D (Q C.1.), innovation strategy in product development (Q C.2.), and contributions of incremental innovation (Q C.6.)

As proposed by the OECD (1997) regarding measuring innovation activity, data for competencies, innovations and R&D activities were collected with respect to the time frame of the past 3 years.

Different market strategies also require different levels of innovation, depending primarily on the specifics of customer demand. To incorporate this effect was included a question on type of production ranging from mass customization, production of standardized series, and production of series specified by the buyer i.e. "made to order" production (Q C.3.) (Duray, 2002).

Business performance was assessed with financial data (section E of the questionnaire) as well as with export activity from the viewpoint of new-market entry (Q C.8.) (Hollensen, 2007; p. 310-315). Ownership characteristics are also known to have an influence on firm performance. This is especially true in the case of transition economies, such as Slovenia, which have witnessed quite recent waves of privatization and an influx of foreign investments, and with this in mind ownership data can provide valuable information (Tether, 2002; Frydman et al. 1999). Question F.1 refers to the relationship between domestic and private ownership of a firm. A request for detailed ownership structure for the year 2006 is made in question F.2.

Questionnaires written in Slovene language were mailed out in June 2007, targeting management-level employees in charge of company R&D in order to diminish the respondent bias. Beforehand, pilot-testing via structured personal interview based on the questionnaire - was carried out in 12 firms. The questionnaire was tested not only for question content, wording, sequence, form and layout, but also for question difficulty and the quality of the provided instructions. Most changes were proposed regarding the wording of the questions and instructions. It also became apparent that firms with multiple product lines could not always provide one uniform response, which can be seen as further justification for the extension of the questionnaire for distinct product lines. In order to diminish the effect of social desirability bias, surveying-by-mail was later employed (Malhotra & Birks, 2003, p. 238).

To increase the response rate, several measures were taken, as summarized in Leong and Austin (2006, p. 191). The aim of the research was clearly defined in the accompanying cover letter. Both envelopes and cover letters were personalized. Follow up calls were made to non-responding firms two weeks after questionnaires had been sent out and a replacement survey was provided on request. Firms were also assured that data would be published only in the form of the aggregate analysis. Moreover, the questionnaire was initially tested by the potential respondents.

6.3 Variables

In devising indicators of competencies I relied predominantly on surveys used in related studies (Song et al., 2005; Wang et al., 2004; Chang, 1996) and questionnaire testing. Research shows that technological competencies (TC) usually encompass three categories: how advanced research and development is (RD_ADVAN), the number of available technological capabilities either within the firm or through strategic partnerships (TECH_CAP_NQ), and how good the company is at predicting technological trends (TECH_TREND_F) (Wang et al., 2004; Eisenhardt & Martin, 2000).

Marketing competencies (MC) capture marketing research as well as other marketing activities (Paul & Peter, 1994). In order to include marketing research and forecast competencies, the indicator "obtaining information about changes in customer preferences and needs" (INFO_CUST) was applied. Competitors' patterns of activities are illustrated with "acquisition of real time information about competitors" (INFO_COMP), customer relationship management with "establishing and managing long-term customer relations" (CUST_RELAT) and supplier relations using the indicator "establishing and managing long-term relations with suppliers" (SUPP_RELAT). Selected indicators to some degree reflect Porter's competitive forces.

Complementary competencies (CC) represent the degree of congruence between technological and marketing competencies. The internal environment is measured with "good transfer of technological and marketing knowledge among business units" (TECH_MRKT_KN), while the indicator "the intensity, quality and extent of research and development knowledge transfer in co-operation with strategic partners" (RD_STP) evaluates dynamic perspective and competence acquisition through strategic partnerships. The efficiency of the economic utilization of technological and marketing resources engaged in product development is assessed through "product development is cost efficient" (RD_COST_EFF), with organizational focus being measured via the indicator "activities of the business units in the corporate strategy of our firm are clearly defined" (ACT_STRAT).

The general extent of innovative performance (IP) was measured by "the number of modified, improved and new products" (NO_CH_PROD) representing new-product variety or level of innovation. Technical performance was added and included using the variable "quality of products" (QUAL_PROD). A number of studies in the operations management literature confirm positive relations between propulsive new product development practices and both product innovation and product quality (Koufteros & Marcoulides, 2006; Clark & Fujimoto, 1991, p. 153-155; Dumaine, 1989). Product quality is furthermore linked to firm performance in that high quality products build brand equity for a firm and lead to the firms in question being in a position to charge price premiums for its products. Studies based on the PIMS database confirm this finding and attribute high-financial measure of revenue to improved market share and profitability due to lower cost (Kroll et al., 1999; Buzzell, 2004). Product quality or technical performance stands for the development and production of products that satisfy customer needs regarding quality and performance (Kim et al., 2005; Hall et al., 1991, p. 25-35).

Furthermore, quality and innovation are considered to be the two most recognized strategic metrics associated with a differentiation strategy (Prajogo et al., 2008; Belohlav, 1993; Hill, 1988).

NPD speed is defined as the pace of activities between idea conception and product implementation. There are several ways in which the NPD cycle speed can positively contribute to revenue and profitability, among them the conferral of first-mover advantage via higher margins, increased market share, the establishing of industry standards and locking up distribution channels. Short NPD cycles are also linked to speedier learning, clearer measures as well as the adoption of performance goals and schedules, lower levels of inventory and working capital and the motivational effects of frequent feedback. Moreover, firms consistently launching new products ahead of the competition also simultaneously build their brand and image (Menon et al., 2002).

The indicator "time needed to develop an improved product" (TIME_IMPR) was applied to determine the effectiveness of improving existing products (incremental innovation). Time refers to the development project lead time and not to the array of products developed, as with the general indicator NO_CH_PROD. Similarly, the effectiveness of new product development referring to radical innovation is measured by "time needed to develop a completely new product" (TIME_NEW) (Chang, 1996). The role of innovativeness of the firm in the industry was represented by the indicator "the firm's substantial contribution to world trends in the industry« (TRENDS). This indicator, TRENDS, makes the assumption of ascribing to market pioneers innovations their competitors find worth imitating. Latent variables of the operational model and their indicators are summarised in Table 2.

There are two ways in which competencies can be measured. The so called inside view proposes measuring competencies in terms of the degree of task performance and qualifications. However, since competencies cannot be observed from the outside, they can be evaluated in relation to competitors (Day, 1994; Prahalad & Hamel, 1990). The latter approach was applied in the present study. The use of this relative self-assessment scale adjusts for intra-sectoral heterogeneity as addressed in the chapter on high- and low- and medium-tech industries. Self-assessment as such is also a widespread practice in firms, allowing them to identify both their strengths and areas in which improvements can be made (Ritchie & Dale, 1999).

The respondents evaluated both competencies and innovative performance on a five-point scale relative to their main competitors and in so doing estimated the competitiveness of their individual competencies within the industry (Song et al., 2005). The scale values were as follows:

- 1 – much worse than the main competitors,
- 2 – somewhat worse than the main competitors,
- 3 – at the level of the main competitors,
- 4 – somewhat better than main competitors,
- 5 – much better than main competitors.

Table 2: Latent variables of the operational model and their indicators

Indicator	Indicator label	Latent variable
Advancement of R&D	RD_ADVAN	Technological competencies (TC)
Number of quality technological capabilities inside the firm or through strategic partnerships	TECH_CAP_NQ	
Prediction of technological trends	TECH_TREND_F	
Establishing and managing long-term customer relations	INFO_CUST	Marketing competencies (MC)
Acquisition of real-time information about competitors	INFO_COMP	
Obtaining information about changes in customer preferences and needs	CUST_RELAT	
Establishing and managing long-term relations with suppliers	SUPP_RELAT	
Good transfer of technological and marketing knowledge among business units	TECH_MRKT_KN	Complementary competencies (CC)
The intensity, quality and extent of research and development knowledge transfer in co-operation with strategic partners	RD_STP	
Cost efficiency of product development	RD_COST_EFF	
Clearly defined activities of business units in the corporate strategy of our firm	ACT_STRAT	
Number of modified, improved and completely new products in period 2004-2006	N_CH_PROD	Innovative performance (IP)
Time needed to develop an improved product	TIME_IMPR	
Time needed to develop a new generation product	TIME_NEW	
Contribution of the firm to industry trends	TRENDS	
Quality of products	QUAL_PROD	

In order to assess firm innovativeness, firms were asked to evaluate the extent to which they were pursuing strategies of innovation and imitation on a 5-point scale with the following categories:

- 1 – only imitation,
- 2 – predominantly imitation,
- 3 – balanced,
- 4 – predominantly innovation,
- 5 – only innovation.

Measures ROA and ROE were included as indicators of profitability and, thus, of integrated business performance (BP). Data from actual financial statements were used. Business performance is measured in the model by the calculated average ROA and ROE during the three year period 2004-2006, i.e. the same period for which the firms were asked to evaluate their innovative performance. ROA measures management's ability and efficiency in issuing the firm's assets to generate profits (White et al., 2003, p. 134-135). ROE, on the other hand, reports on the return on total stockholder equity.

The success of innovations – as mirrored in the price premium the firm is able to attain for its new products on the market - was assessed by the indicator value added (ADD_VAL) which, in accounting sense, represents the difference between revenues and costs of goods/services sold/provided (Treacy & Wiersima, 1993). Respondents ranked this indicator in the same way as they did competencies. While cost-efficiency of the firm denotes that efficiency the company tries to increase by exploiting all of the resources at its disposal (Ravald & Grönroos, 1996), it was included as a self-assessment indicator of the overall performance of the firm (BP_COST_EFF).

Four different indicators were applied to each category of the environmental turbulence (Song et al., 2005; Wang et al., 2004; Calantone et al., 2003). In the case of technological turbulence, the measured elements were: the speed of change in technology, opportunities arising due to new technologies, the ability to predict technological change, and the degree of technological change in the industry. Questions regarding market turbulence referred to market uncertainty, the predictability of changes in demand, the predictability of competitors' activities, and competition intensity. Indicators of environmental turbulence were evaluated on a five-point Likert scale. Environmental turbulence reflects, to a great extent, the specifics of the industries in which firms operate.

6.4 Data

The population targeted in the survey was obtained from the database of legal entities provided by the Agency of the Republic of Slovenia for Public Legal Records and Related Services (slo. Agencija Republike Slovenije za javnopravne evidence in storitve – AJPES). Changes in accounting standards somewhat affected the collection of data for the fiscal year 2006 as data on exports and employee numbers were no longer available in the commercial database.

Companies were selected according to the CPA 2002 classification (Statistical Classification of Products by Activity in the European Economic Activity) provided by Eurostat. Included companies were those with products under code D (manufactured products) without codes ending with 9 (xx.xx.9) that refer to product-related industrial services. For problems arising from product finishing industries such as production of clothing items, several further product codes were excluded. This is to avoid the potential confusions stemming from aligning the design function in these companies with the definition of the traditional R&D function and related activities in manufacturing firms. Other product groups were selected as presented in 0.

Furthermore, only those companies that had been registered prior to 2002 and had been operating throughout the whole period 2002-2006 were included. The population has been additionally narrowed down to medium-sized and large companies with established business functions. The target population of companies thus consisted of 187 medium-sized and 194 large companies; in total, 381 companies.

The size of firms was adopted from the AJPES database in accordance with the 55th article of the Companies Act (2006; slo. Zakon o gospodarskih družbah – ZGD-1). The definition adheres to criteria regarding the average number of employees in a financial year, net sales income, and the value of assets. Each size category is defined by meeting two of the criteria, as follows:

- Micro company:
  - average number of employees in a financial year does not exceed 10,
  - net sales income does not exceed 2,000,000 EUR, and
  - value of assets does not exceed 2,000,000 EUR.
- Small company:
  - average number of employees in a financial year does not exceed 50,
  - net sales income does not exceed 7,300,000 EUR, and
  - value of assets does not exceed 3,650,000 EUR.
- Medium-sized company:
  - average number of employees in a financial year does not exceed 250,
  - net sales income does not exceed 29,200,000 EUR, and
  - value of assets does not exceed 14,600,000 EUR.

Medium-sized company criteria at the same time define the lower threshold applying to large companies.

In total, 53 companies returned valid questionnaires yielding a 13.9% response rate. Companies were asked to provide data for individual product lines where applicable. Nine companies gave responses for more than one product line thus providing a total sample of 70 observations. As a result of further analysis, 3 companies with 5 product lines in total were excluded, due to consecutive negative EBIT results.

Table 3: CPA 2002 product groups selected for the target population of manufacturing firms (medium-sized and large) including number of respondents

Code		Products by activity	No. of firms in population	No. of responding firms	No. of reported product lines**
DA	15	Food products and beverages	39	1	1
	16	Tobacco products			
DB	17	Textiles *	9	2 (4)***	2 (4)***
	18	Wearing apparel; furs *			
DC	19	Leather and leather products *	1	0	0
DD	20	Wood and products of wood and cork; except furniture; manufacture of articles of straw and plaiting materials	15	0 (1)***	0 (3)***
DE	21	Pulp, paper and paper products	13	1	3
	22	Printed matter and recorded media *			
DF	23	Coke, refined petroleum products and nuclear fuel	0	0	0
DG	24	Chemicals, chemical products and man-made fibers	33	10	14
DH	25	Rubber and plastic products	36	3	3
DI	26	Other non-metallic mineral products	25	1	1
DJ	27	Basic metals	56	6	6
	28	Fabricated metal products, except machinery and equipment			
DK	29	Machinery and equipment not earlier classified	53	4	7
DL	30	Office machinery and computers	51	17	21
	31	Electrical machinery and apparatus not earlier classified			
	32	Radio, TV, communication equipment and apparatus			
	33	Medical, precision and optical instruments; watches and clocks			
DM	34	Motor vehicles, trailers and semi-trailers	17	2	2
	35	Other transport equipment			
DN	36	Furniture; other manufactured goods not earlier classified	33	3	5
	37	Secondary raw materials			
		Total:	**381**	**53**	**70**

Note:
* Excluded were: 17.3-17.7: textile finished products; 18: all; 19.2-19.3: luggage, handbags and the like; saddlery and harness, footwear; 22: all; and all industrial services with codes xx.xx.9.
** Reported product lines with distinctive competencies.
*** Numbers in parentheses denote the number of responses obtained, including those observations that were excluded as outliers for the analysis (consecutive years of negative EBIT).
Source: AJPES, 2007 and survey data.

6.4.1 General company data

The majority of firms in the sample are large firms (76%). As of the year 2006 most firms (86%) have been present in their respective industries for more than 30 years even though most changed their legal status during the process of privatization which took place in the 1990's, but remained within the same industry. Only 2 companies have between 5 and 10 years of experience, while 5 companies have between 10 and 20 years of experience in the industry. In total, 72% of the companies belong to a formal group of firms with interrelated ownership. 78% of the companies in the sample have majority domestic ownership, 20% foreign and there is also one company with the relatively uncommon status of being of equal-share domestic and foreign ownership.

While only 6 companies in the sample have only one distinct product line, 27 companies have 2 or 3 and 17 companies have 4 or more. The sales generated in 2006 by the firms in the sample ranged from 3,624,000 EUR to 733,308,000 EUR, with the average sales amounting to 79,199,000 EUR. The average annual growth of sales in the 5 year period from 2002 to 2006 inclusive was 8.80%, with respect to which it is important to note that some firms witnessed negative sales growth rates. While 643.68 was the average annual number of employees of the 53 firms in 2005, the median was only 275.87, with the average gross wage being 1,552 EUR. The data are presented in Table 4. It can be concluded that the firms are export-oriented. In 2005 the firms generated, on average, 71.86% of their sales abroad.

The mean value of the reported R&D expenditure in 2006 is 4.48%, measured as a percentage of total sales. Companies do keep track of their R&D expenditure, a practice that was, until recently, strongly encouraged by tax conditions which were favourable in terms of income tax benefits. It is quite different when advertising expenditure is considered. Many companies do not yet account for this expenditure in a separate category and are therefore unable to provide the data. Those that do have a better understanding of their advertising expenditure in accounting terms, however, reported very low values; on average, below one percent.

It is encouraging to note that the firms generate the vast majority of their sales through products branded as their own. Not only is this important as it allows the company to enhance its brand's recognition, but also because own-brands make higher price mark-ups possible. This can be seen from the added value which is calculated as the mark-up on costs of goods sold that is reflected in the prices of these products. On average, the firms thus managed to earn a 40.46% gross margin on their products.

On average, the firms replace two thirds of their product portfolio, measured as a percentage of sales, within 3 years. These data point both to short product lifecycles and to intense competition through R&D as well as design.

Table 4: General firm data

Data for year 2006	Mean	Median	Std. Dev.	Min	Max
Sales (000 EUR)	79,199	29,346	12,524	3,624	733,308
Sales growth during 2002-2006 (annual average)	8.80%	9.25%	6.76%	-10.94%	26.10%
No. of employees (2005)	643.68	275.87	974.41	43.63	5,673.66
Gross wage per employee 2005 (EUR)	1,552	1,512	449	930	3,180
Share of export in total sales (2005)	71.86%	76.58%	23.66%	1.40%	98.80%
R&D expenditure as % of sales	4.48%	3.00%	4.01%	0.00%	17.00%
Advertising expenditure as % of sales	0.98%	0.85%	1.03%	0.00%	4.50%
Added value*	40.46%	36.42%	19.97%	7.51%	105.60%
Sales under own brand (%)	82.02%	100%	30.38%	0.00%	100%
Incremental and radical innovation as a share of sales during the past 3 years (%)	66.40%	87.50%	37.04%	0.00%	100%

Note: * Added value calculated as the difference between sales and costs of goods sold, relative to costs of goods sold (multiplied by 100 to obtain %). All same-year data.

Source: Survey data and own calculations.

Table 5: Ownership structure (% of total)

Data for 2006	Mean	Median	Std. Dev.	Min	Max
State funds	7.05	0.00	16.51	0.00	70.50
Investment funds	7.48	0.00	15.97	0.00	61.83
Other companies	54.34	68.48	44.73	0.00	100.00
Banks	1.25	0.00	3.63	0.00	16.00
Minority owners	3.36	0.00	8.23	0.00	49.00
State of Republic of Slovenia and municipalities	0.00	0.00	0.00	0.00	0.00
Employee ownership	0.00	0.00	0.00	0.00	0.00
Management	4.82	0.00	14.09	0.00	80.00
Ex-employees, retired employees, relatives	21.01	0.00	38.42	0.00	100.00
Non-realized internal buyout	0.91	0.00	2.65	0.00	15.00
Other	0.97	0.00	3.94	0.00	23.50

Source: Survey data and own calculations.

The results concerning ownership structure (Table 5) reveal that the most common owners of the companies in the sample are other companies (on average 54.3%). This is in accordance with the fact that most of these companies belong to formal groups of companies. The second most important category of owners appear to be ex-employees, retired employees and their relatives (21.0%), followed by state funds and investment funds (both with average ownership shares roughly at 7%). At slightly less than 5% ownership, management has a rather small share.

6.4.2 R&D activities and the production function

Companies in the sample rely predominantly on internal R&D activities. 33.8% of the 65 product lines in the sample carry out only internal R&D (Table 6). 49.2% of product lines are the subjects of joint R&D, wherein internal R&D activities are dominant. Balanced (7.7%) and prevailing external R&D are somewhat rarer (9.2%), with no reports whatsoever of external R&D alone. In those cases where companies cooperate at the level of R&D, 36.9% product lines recognize in-house R&D to be the key source of added value, followed by equal added value provided by both types of research input. (23.1%) Only 6.2% of product lines engaging in joint R&D find external R&D to contribute more to the added value of innovations.

Table 6: R&D function

Replies provided for all distinct product lines specified	Answer categories	No. of product lines (total 65)
Internal or external R&D function	Only internal	33.8%
	Internal prevailing	49.2%
	Balanced	7.7%
	External prevailing	9.2%
	Only external	0.0%
Added value of collaborative innovation (in total 43 product lines)	Internal grater	36.9%
	Equal	23.1%
	External greater	6.2%
Innovation and imitation in R&D	Only imitation	6.2%
	Imitation prevailing	40.0%
	Balanced	29.2%
	Innovation prevailing	20.0%
	Only innovation	4.6%

Source: Survey data and own calculations.

Although imitation prevails as the predominant strategy in R&D, innovation is nonetheless strong as well. Of all observed product lines, 6.2% depend solely on imitation and 40.0% depend largely on imitation. In 29.2% of the cases both imitation and strategy of innovation are employed in a balanced way. 4.6% of product lines solely depend on innovation in their R&D activities.

Firms rated each specific innovation goal in its R&D activities on a 5-point scale, with value 1 meaning "not important" and 5 "very important" (Table 7). All of the innovation goals proved to be of significant importance. The highest average value was ascribed to better company image (4.25), followed by improved appearance (4.15), this last pointing to the role of design. Improved product functionality received the third highest score, while lower production costs and improved product use were assigned the lowest values. The relatively lower importance of improved product use could be explained by many products being intermediate products that are, as such, developed to fit the requirements of the end product of which they are to be a constituent part.

Table 7: Innovation goals

Innovation goal	Mean	Median	Std. Dev.	Min	Max
Improved product use	3.20	3.00	1.31	1.00	5.00
Improved product functionality	3.78	4.00	1.27	1.00	5.00
Lower production costs for your company	3.40	3.00	1.07	1.00	5.00
Improved appearance	4.15	5.00	1.06	1.00	5.00
Better company image	4.25	5.00	1.02	1.00	5.00

Source: Survey data and own calculations.

The most widespread type of production is the production of standardized series (Table 8). On average, 48.0% of total product volume for the product lines in question is based on standardized series. This is followed by the production of series specified by the buyer, where the average percent of volume produced is 33.1%. The other two options are considerably less represented. Out of 65 product lines, 58.5% make no use whatsoever of customized production and only 24.6% employ and implement mass customization.

Table 8: Types of production as share of total quantities produced (%)

Data for 2006	Mean	Median	Std. Dev.	Min	Max
Customized production	10.51	0.00	25.14	0.00	100.00
Production of series specified by the buyer	33.14	20.00	36.25	0.00	100.00
Production of standardized series	48.03	50.00	40.37	0.00	100.00
Mass customization	8.32	0.00	22.77	0.00	100.00

Source: Survey data and own calculations.

6.4.3 Aggregate R&D company data

Research efforts regarding improvements to existing products and technologies constitute, on average, the largest proportion of the R&D expenditure of the companies in the featured sample (36.1%) (Table 9). The second most dominant R&D

expenditure category is the development of new-generation products, which is, compared to the incremental innovation of the previous category, connected with more risk and higher costs. Expenditures relating to the development of new production methods and processes, on average, amount to 17.7%. Basic research and laboratory activities are very scarce (8.0% and 7.1% respectively). It can be observed that incremental innovation take up the largest portion of the R&D expenditure of the firms comprising the sample.

Regarding R&D expenditure, it is important to keep in mind that R&D expenditure may not provide a complete picture of those companies that engaged in a considerable amount of R&D through strategic partnership; in such cases, R&D activities may be outsourced to a partner that subsequently becomes an exclusive supplier to the firm.

Table 9: R&D expenditure structure as a share of total R&D expenditure during 2004-2006 (%)

Period 2004-2006	Mean	Median	Std. Dev.	Min	Max
Basic research	8.00	5.00	10.57	0.00	50.00
Research for improving existing products and technologies	36.14	30.00	27.11	0.00	100.00
Development of new generation products	27.18	20.00	23.53	0.00	90.00
Development of new production methods and processes	17.69	10.00	22.90	0.00	100.00
Laboratory activities	7.15	5.00	9.51	0.00	40.00

Source: Survey data and own calculations.

Table 10, which presents the structure of the financing sources of R&D expenditure, shows that internal funds are the predominant source (84.7%). Already during the pilot testing of the questionnaire it emerged that internal sources are the most consistent, while others fluctuate depending on specific projects and are, therefore, temporary. Loans, state funding and funding from the European Union amount to several percent each. The level of joint investment with both domestic and foreign partners is very low (0.5% and 1.8%), something which also holds for funding through universities and research institutions (0.4%). All of these three sources are linked to collaborative R&D efforts.

Table 10: Structure of financing sources of R&D as a share of total R&D expenditure during 2004-2006 (%)

Period 2004-2006	Mean	Median	Std. Dev.	Min	Max
Internal sources	84.74	95.50	24.46	0.00	100.00
Loans	4.90	0.00	15.83	0.00	90.00
Joint investment with domestic partners	0.51	0.00	1.75	0.00	10.00
Joint investment with foreign industrial partners	1.85	0.00	7.64	0.00	45.00
Universities and research institutions	0.45	0.00	1.71	0.00	10.00
State funding	3.67	0.00	8.15	0.00	44.00
EU	2.06	0.00	7.20	0.00	40.00

Source: Survey data and own calculations.

6.5 Innovative performance based clustering

In this section the aim is to identify groups of distinct product lines that share similar characteristics with respect to innovative performance and underlying competencies. As presented in the literature review, technology leaders and followers develop different sets of competencies on which they build their competitive advantage. In order to obtain segments of firms' product lines based on their innovative performance, a clustering procedure on the variables N_CH_PROD and QUAL_PROD reflecting innovative performance will be carried out. To organize observed cases into these relatively homogenous groups, techniques of cluster analysis or data segmentation are applied. While objects within the same group – cluster – share similarities, they tend to be different compared to objects within other clusters. Comparisons of clusters not only provide an insight into such differences but thereby also enable an understanding of their own characteristics. As firms, large ones in particular, try to take advantage of synergies and economies of scale and scope, many diversify into different businesses. A distinction is thus made between specific businesses or product lines within the company, as identified by the respondents.

The analysis begins with the hierarchical method, which divides clustering data into subsets by finding clusters which succeed those already established. According to Formann, the minimal sample size should equal 2^k (Dolničar, 2003), where k is the number of variables in the segmentation base, or preferably $5*2^k$. In the analysis presented the minimum sample size required is thus $5*2^2 = 20$ (< 65).

Agglomerative clustering presents a "bottom-up" approach by grouping objects into bigger and bigger clusters. The opposite is divisive clustering, a "top-down" approach, which begins with objects grouped as a single cluster and subsequently divides and subdivides each object in a separate cluster.

Agglomerative hierarchical clustering procedure generates a partition sequence of the data of the following form: $P_n, P_{n-1}, ..., P_1$. The first partition P_n is composed of n single object "clusters", while the last single groups partition P_1 contains all n cases. The method joins at each step the two closest and most similar clusters. Agglomerative techniques vary in how they define distance (similarity) between clusters. The average linkage method and Ward's procedure have been shown to be superior to others (Malhotra & Birks, 2003, p. 602-603). With average group linkage the formed groups are represented by their mean values for each variable – their mean vector. Distances between groups are defined in terms of the distance between two such mean vectors. Ward (1963) developed a clustering procedure that seeks to form partitions in a way that minimizes "information loss with each grouping." It is a method in which the squared Euclidean distance to the cluster means is minimized, and calculates the distance between clusters according to the following equation:

$$d(C_i \cup C_j, C_k) = \frac{(n_i + n_j)n_k}{(n_i + n_j + n_k)} d^2(T_{i,j}, T_k) \tag{15}$$

Where clusters C_i and C_j are the closest and therefore joined $C_i \cup C_j$. The distance between this new group and C_k is then calculated. The numbers of objects belonging to a specific cluster are denoted by n_i, n_j and n_k respectively. The distance between cluster centroids $T_{i,j}$ and T_k is denoted by d. A graphical representation of both methods is provided in Figure 5.

Figure 5: Graphical representation of average group linkage and Ward's hierarchical clustering method

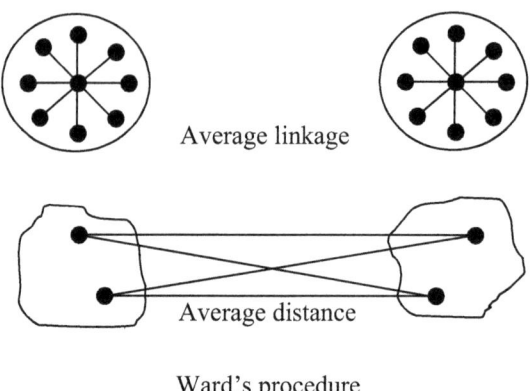

Average linkage

Average distance

Ward's procedure

Source: Malhotra and Birks, 2003, p. 602.

To identify final clusters a two step methodology was used (Ferligoj, 1989, p. 88). Ward's procedure was applied first and a dendrogram was obtained (Appendix D). This is a graphical representation of fusions made at each successive stage of partitioning. 5 observations belonging to firms reporting consecutive losses during the observed period were removed as outliers, thus yielding a sample of 65 observations.

According to the two step clstering methodology, non-hierarchical methods are applied next in order to improve the classification if necessary. MacQueen's K-means method was employed by calculating the centroids for the 3 previously defined clusters as seeds. Centroids are calculated as the average values of variables for each separate cluster. The method organizes observations into those clusters whose centroids are the closest. Since the conclusions of this method depend on the order of cases in the data set, they were first ordered according to their classification as yielded by the hierarchical method. In the ensuing steps the method repeatedly calculates the centroids of the new clusters in case any objects have been moved. As seen in Table 11, only one iteration was performed.

The K-means method classified 1 object out of 65 into the clusters differently from the hierarchical Ward's procedure (classification table in Appendix E). Convergence was achieved due to a lack of change in cluster centres. The current iteration is 1, with the minimum distance between initial centres being 0.751. Thus, hierarchical clustering had already produced a good solution.

Table 11: K-means method iteration history report

Iteration	Change in cluster centers		
	1	2	3
1	0.000	0.000	0.000

Three distinct segments were identified which were further compared in terms of competencies, in order to gain a deeper understanding of the differences obtaining between them. They were compared using ANOVA and "post-hoc Duncan test" (equal variances assumed), P<0,05 (see Table 12).

In Table 12 pluses ([+] in the table) marked next to the average values of variables for each segment denote whether the differences between segments are statistically significant. If they are not, segments are given the same number of pluses. If differences are established, segments are given varying numbers of pluses, the one with the most being that with the highest mean value. Turning to the variable N_CH_PROD, we can conclude that there are no statistically significant differences observed between the first and second segments (both denoted by one plus [+]). However, there is statistically significant difference between the first two segments, on the one hand, and the third segment, which is ascribed two pluses [++], on the other. As such third segment is demonstrating on average higher values of variable N_CH_PROD compared to the first two.

The following three segments were identified (Table 12):

- technology followers with weak competencies,
- technology followers with strong competencies, and
- technology leaders.

Based on indicators of innovative performance, it can observed that the first segment - technology followers with weak competencies - introduced the smallest number of new products in the past 3 years (N_CH_PROD) as well as those of the poorest quality relative to their main competitors (both variable scores are below the level of main competitors, which is value 3). Conversely, it is the third segment - technology leaders - that surpasses main competitors according to both indicators (values above 4 – "better than main competitors"). While the second segment is lagging behind in terms of the number of new products introduced in the past 3 years (modified, improved and completely new products combined), it appears to compensate for the lack of new product variety to some extent with the high quality of those new products it does generate. Further implication that we are dealing with technology followers in the case of the first two segments is provided by their predominant strategy being that of imitation (values below 3 – "balanced innovation"), which is technologically less demanding.

Table 12 (1/2): Product lines segments described by innovative performance, competencies and NPD characteristics

Variables		Segments		
		Technology followers - weak	Technology followers - strong	Technology leaders
No. of product lines		25	19	21
No. of different companies		21	16	20
Innovative performance (IP)				
Number of modified, improved and completely new products in period 2004-2006	N_CH_PROD	2.84 +	2.89 +	4.24 ++
Quality of products	QUAL_PROD	2.96 +	4.21 ++	4.24 ++
Technological competencies (TC)				
Advancement of R&D	RD_ADVAN	2.84 +	3.16 +	3.86 ++
Number of quality technological capabilities inside the firm or through strategic partnerships	TECH_CAP_NQ	2.72 +	3.32 ++	4.10 +++
Prediction of technological trends	TECH_TREND_F	2.68 +	3.00 +	3.95 ++
Marketing competencies (MC)				
Obtaining information about changes in customer preferences and needs	INFO_CUST	2.92 +	3.26 +	3.95 ++
Acquisition of real time information about competitors	INFO_COMP	3.00 +	3.16 +	3.29 +
Establishing and managing long-term customer relations	CUST_RELAT	3.32 +	3.79 ++	4.10 ++
Establishing and managing long-term relations with suppliers	SUPP_RELAT	2.92 +	3.58 ++	3.67 ++

Table 12 (2/2): Product lines segments described by innovative performance, com-petencies and NPD characteristics

Variables		Segments		
		Technology followers - weak	Technology followers - strong	Technology leaders
No. of product lines		25	19	21
No. of different companies		21	16	20
Complementary competencies (CC)				
Good transfer of technological and marketing knowledge among business units	TECH_MRKT_KN	2.80 +	3.32 ++	3.52 ++
The intensity, quality and extent of R&D knowledge transfer in co-operation with strategic partners	RD_STP	2.48 +	3.00 +	3.57 ++
Cost-efficiency of product development	RD_COST_EFF	2.84 +	3.37 ++	3.52 ++
Clearly defined activities of business units in the corporate strategy of our firm	ACT_STRAT	2.88 +	3.58 ++	3.62 ++
New product development				
Time needed to develop an improved product	TIME_IMPR	2.76 +	3.21 ++	3.76 +++
Time needed to develop a new generation product	TIME_NEW	2.48 +	2.63 +	3.71 ++
Contribution of the firm to industry trends	TRENDS	2.44 +	2.47 +	3.24 ++
Imitation VS innovation strategy		2.32 +	2.74 +	3.33 ++

Note: For each variable a segment is described by a mean value (except number of product lines and number of firms in the sample). Pluses denote segments with statistically significant differences. Applied was ANOVA, "post-hoc Duncan test", $P<0.05$.

There is a distinct gap between the first and the third segment when analyzing all three groups of competencies, the first having weaker competencies than main competitors and the third more highly developed ones (statistically significant differences denoted by different numbers of pluses [+]). The only exception to this general rule is found in connection with the acquisition of real time information on competitors (INFO_COMP) among marketing competencies. More about this competence is explained in the continuation.

When addressing technological competencies separately, technology leaders surpass both segments of followers with regards to all three competencies (RD_ADVAN, TECH_CAP_NQ and TECH_TREND_F). The one technological competence that sets apart both segments of technology followers is TECH_CAP_NQ at which technology followers with strong competencies reach the level of their main competitors. This competence regarding the number of quality capabilities which can be accessed either internally or externally is also the one in which technology leaders scored best within technological competencies (value 4.10 – "better than main competitors").

The marketing competence that sets technology leaders apart from technology followers with strong competencies is INFO_CUST. No statistically significant differences can be observed between leaders and followers with strong competencies with respect to establishing and managing long-term relationships with customers and suppliers (CUST_REALT, SUPP_RELAT). Nevertheless, it is in terms of these two competencies that the segment of followers with weak competencies lags furthest behind. There are however no differences between the segments in terms of their competence in acquiring real-time information about competitors (INFO_COMP), all reaching the level of their main competitors. It appears this kind of information is available to all 3 segments in a similar way. Access to real time information about competitors can thus no longer be a differentiating factor in terms of creating and sustaining competitive advantage. It is important to note that marketing competencies as a whole appear to be the most competitive group of competencies for the segment of followers with weak competencies reaching values close to 3.

Among complementary competencies, only RD_STP sets technology leaders apart from followers with strong competencies. This competence of knowledge transfer with strategic partners is also somewhat closely related to the technological competence TECH_CAP_NQ (number of quality capabilities which can be accessed either internally or externally) in which followers with strong competencies also trail the leader. Not only do strategic technologic partnerships have the potential to benefit TECH_CAP_NQ, but also RD_ADVAN due to the availability of new knowledge. While both segments have a clear and well defined strategy, a cost efficient R&D and efficient transfer of technological and marketing knowledge, followers with strong competencies share the same level of competitiveness in RD_STP with the weakest segment.

Technology leaders perform very favourably regarding NPD lead times, also making greater contributions to industry trends and relying more on innovation than imitation. The segment of technology followers with strong competencies is also competitive when it comes to lead times in developing improved products, although not to the extent of technology leaders. Unlike technology leaders, both follower segments are expected neither to report favourable lead times in developing completely new products, nor to contribute substantially to trends in the industry. Along these lines, followers rely predominantly on imitation.

This part of the analysis already provides partial support for Hypothesis 1, in the form of the statistically significant differences in competencies found among firm segments that had been grouped based on their innovative performance. Differences in innovative performance therefore appear to be linked to differences in competencies. Further evidence to support this hypothesis is presented in the analyses that follow.

Three segments of firms were established, including one group of technology leaders displaying strong innovative performance and competencies developed beyond the level of their main competitors. Two different segments of technology followers with weaker innovative performance were also observed, one of which maintains competitive position through the possession of competencies at the level of competitors, the other clearly lagging behind. The most significant gap between competitive technology followers with strong competencies and technology leaders was observed in technological competencies while they maintain relatively high levels of marketing and complementary competencies. This confirms Hypothesis 2. The differences between the two segments of technology followers, both of which still engage to some extent in innovative activity, speak in favour of Hypothesis 3. Technology followers with strong competencies at the same time exhibit better business performance than followers with weak competencies. The innovative activity of both segments relies most heavily on incremental innovation and imitation as implied in Hypothesis 4.

6.6 Structural models

Structural equation modelling (SEM) is a collection of statistical techniques that facilitate the examination of a set of relationships between one or more independent and dependent variables. To test the hypotheses, Partial Least Squares (PLS) approach to structural modelling was employed due to limitations in sample size as this method makes minimal demands in terms of measurement scales, sample size and residual distributions. It can be used for both establishing theory and for confirmation purposes or theory testing.

Unlike some of the well known factor-based, covariance fitting approaches for latent structural modelling, among them LISREL, EQS and AMOS, PLS is component based. Therefore it avoids the problems of inadmissible solution and factor indeterminacy (Fornell & Bookstein, 1982). With factor-based covariance fitting approach, the indeterminacy of factor score estimations can lead to a loss of predictive accuracy, which constitutes a problem in the case of theory development. This approach makes use of covariance based full-information estimation methods, among them Maximum Likelihood or Generalized Least Squares. Chin et al. (2003) suggest the PLS approach is in many cases more suitable for application and prediction purposes. It is namely assumed that all the measured variance in useful variance is to be explained. Latent variables are estimated as exact linear combinations of the observed measures. By avoiding the indeterminacy problem it provides an exact definition of component scores. It uses the iterative estimation technique (Wold, 1981) and provides a general model encompassing techniques such as canonical correlation, redundancy analysis, multiple regression, multivariate analysis of variance, and principal components. The iterative algorithm generally consists of a series of ordinary least squares analyses, such that identification is not a problem for recursive models. At the same time, it does not presume any distributional form for measured variables.

Regarding sample size, a strong rule of thumb defines it as being equal to or larger of the following (Chin et al., 2003):

- ten times the scale of the largest number of formative (causal) indicators (this does not apply to the use of reflective indicators), or
- ten times the largest number of structural paths directed at a particular construct in the structural model.

A weaker rule of thumb suggests using a multiplier of five instead of ten. While PLS is regarded as a better suited option for explaining complex relationships (Fornell et al., 1990), it is argued that PLS is less appropriate for confirmatory analysis, being primarily intended for causal-predictive analysis in situations of high complexity and low theoretical information (Wold, 1982). In this research SmartPLS 2.0 (beta) software (Ringle et al., 2005) was used to perform the PLS analysis of structural models.

A structural model requires two types of models; namely, (a) the measurement model (so-called outer model) that connects the manifest variables (indicators, items) to the latent variables (constructs), and (b) the structural model (inner model) that connects the latent variables with one another. So as to assess the measurement model, the types of relationship between the latent constructs and the indicators have to be specified first. The reflective approach was applied due to the manifest variables or indicators in the model being considered to reflect their latent variables (Tenenhaus et al., 2005).

6.7 Structural models of competencies and innovative performance

The proposed model of competencies and innovative performance was first assessed for the sample of 65 product lines. Manifest and latent variables along with their labels used in the analysis are presented in Table 13. The proposed model is graphically presented in Figure 6 and consists of four latent variables (constructs); that is, three groups of competencies (TC, MC and CC) and innovative performance IP, all of which are represented by circles. The 13 indicators or manifest variables are represented by square boxes. With 3 structural paths the sample size requirement for the reflective model is met with N = 65 being larger than 10*3 = 30.

Firstly, 4 distinct models of innovative performance were analyzed by substituting the second indicator of innovative performance (IP) with 4 different variables. The baseline model measures IP with indicators NO_CH_PROD and QUAL_PROD. New product variety as a result of a firm's innovative activity is accounted for by the variable NO_CH_PROD. The technical dimension of new product performance is measured by QUAL_PROD. In order to analyze the differences between competencies relating to superiority in R&D activities regarding (a) incremental innovation captured in improved products, and (b) radical innovation captured in new generations of products, the general indicator of the construct innovative performance was substituted accordingly. To account for incremental innovation a replacement indicator "time needed to develop an improved product" (TIME_IMPR) was introduced, for radical innovation indicator "time needed to develop a new generation product" (TIME_NEW), and for the trend-setting role of a firm in the industry, the indicator "contribution of the firm to industry trends" (TRENDS). All indicators and their corresponding latent variables are listed in Table 13.

The different models of innovative performance were initially checked for internal consistency reliability, convergent validity and discriminant validity in order to establish the adequacy of latent variables with respect to capturing their corresponding manifest variables (steps proposed by Anderson and Gebring (1988)). The proposed models were assessed for the sample of 65 product lines of 50 firms.

Table 13: Latent variables and their indicators

Indicator	Indicator label	Mean	St. Dev.	Latent variable
Advancement of R&D	RD_ADVAN	3.22	0.932	Technological competencies (TC)
Number of quality technological capabilities inside the firm or through strategic partnerships	TECH_CAP_NQ	3.32	0.935	
Prediction of technological trends	TECH_TREND_F	3.18	0.896	
Obtaining information about changes in customer preferences and needs	INFO_CUST	3.74	0.828	Marketing competencies (MC)
Acquisition of real time information about competitors	INFO_COMP	3.20	0.670	
Establishing and managing long-term customer relations	CUST_RELAT	3.40	0.857	
Establishing and managing long-term relations with suppliers	SUPP_RELAT	3.36	0.722	
Good transfer of technological and marketing knowledge among business units	TECH_MRKT_KN	3.20	0.756	Complementary competencies (CC)
The intensity, quality and extent of research and development knowledge transfer in co-operation with strategic partners	RD_STP	2.98	1.059	
Cost-efficiency of product development	RD_COST_EFF	3.24	0.797	
Clearly defined activities of business units in the corporate strategy of our firm	ACT_STRAT	3.28	0.809	
Number of modified, improved and completely new products in period 2004-2006	N_CH_PROD	3.36	0.921	Innovative performance (IP)
Time needed to develop an improved product	TIME_IMPR	3.30	0.839	
Time needed to develop a new generation product	TIME_NEW	3.00	1.069	
Contribution of the firm to industry trends	TRENDS	2.82	1.063	
Quality of products	QUAL_PROD	3.70	0.707	

Analysis of the baseline model of innovative performance shows that internal consistency reliability can be confirmed since the values of composite reliability for all constructs exceed the threshold of 0.70, the minimum value being 0.7869 (Table 14).

Table 14: Composite reliability, correlation matrix and the square roots of AVE

	Composite reliability	TC	MC	CC	IP
TC	0.9197	*0.8903*			
MC	0.8438	0.6542	*0.7613*		
CC	0.8340	0.6583	0.7260	*0.7474*	
IP	0.7869	0.7273	0.6885	0.6861	*0.8054*

Note: The square roots of AVE are in the diagonal in italics. Below the diagonal are correlation coefficients.

Table 15 shows only those cross loadings with values larger than the mean of the absolute values, 0.6027. The suggested cut-off for factor loadings is 0.60 (Hatcher, 1994). The minimum value of proposed indicators in the observed model is 0.6134. All latent variables are well correlated with their own manifest variables. Thus, manifest variables adequately describe their latent variables and are, in so doing, validated, thus demonstrating the convergent validity. Furthermore, the average variance extracted (AVE) is higher than 0.50 for each construct (see square roots of AVE in Table 14). This criterion guarantees that in the measurement of a construct there is more valid variance explained than error (Fornell & Cha, 1994).

Fornell and Cha (1994) also provided the criterion for discriminant validity according to which the square root of AVE of each latent variable should be higher than all of its correlations with other latent variables in the model. The square root of AVE for each construct is stated in the diagonal in Table 14 and, as can be seen, they are higher than the correlation coefficients directly below them. This indicates that the latent variables in the proposed model are both conceptually and empirically distinct from each other.

Table 15: Cross loadings between manifest and latent variables

Indicators	TC	MC	CC	IP
RD_ADVAN	**0.8481** (19.396)			
TECH_CAP_NQ	**0.9076** (44.542)			0.7019
TECH_TREND_F	**0.9139** (42.422)	0.6286	0.6145	0.6675
INFO_CUST	0.6154	**0.8452** (26.603)	0.6227	0.6738
INFO_COMP		**0.6134** (5.539)		
CUST_RELAT		**0.8740** (27.265)	0.6248	
SUPP_RELAT		**0.6812** (9.183)		
TECH_MRKT_KN			**0.7955** (10.885)	
RD_STP	0.6534		**0.7589** (17.404)	
RD_COST_EFF			**0.6329** (5.227)	
ACT_STRAT		0.6908	**0.7910** (2.054)	
N_CH_PROD	0.6427			**0.8187** (15.836)
QUAL_PROD				**0.7919** (12.611)

Note: T-values are stated in parentheses for those indicators that belong to a designated latent variable in the model. All significant at P<0.001.

For the other three models internal consistency reliability was also confirmed as the values of composite reliability for all constructs in all four models exceed the stated threshold of 0.70. The values of cross loadings for proposed indicators in the observed models are above the cut-off point of 0.60. All latent variables are well correlated with their indicators. Furthermore, the values of average variance extracted (AVE) are above 0.50 for each construct. Requirements of convergent and discriminant validity are thus also satisfied. Detailed tables are included in the Appendices F, G and H.

Since PLS does not make any distributional assumptions, a bootstrapping method of resampling with replacement was applied, with standard errors being computed on the basis of 500 bootstrapping runs and 65 cases, corresponding to the number of observation units. This was in line with the procedure proposed by Andrews and Buchinsky (2000).

Results for the path coefficients of the baseline model (Figure 6 and Table 16) show that technological competencies have the highest path coefficient and, therefore, the biggest impact on innovative performance. They are followed by marketing competencies and complementary competencies. This finding is similar to that of Jeong et al. (2006), who claim that the technological orientation of firms has a greater impact on technical performance and profitability than customer orientation, however, the latter is more crucial from the viewpoint of customer acceptance of new products. In order to facilitate the coordination of both groups of competencies, complementary competencies are necessary.

By comparing the path coefficients of the four models, it can be observed that the incremental innovation model with the indicator TIME_IMPR yields results that are approximately the same as those provided by the baseline model, with the exception being that complementary competencies play a more important role than marketing competencies. This result shows that the efficiency of new-product development processes relies to a greater extent on competencies of a technological nature than on those of the marketing type. It is in line with the finding of Swink and Song (2007) that integration of technological and marketing knowledge can prolong the technological development stage of a new-product development process. This is even more evident in development endeavours which are technologically more demanding. These are, namely, the development of new-generation products (TIME_NEW) and the setting of trends (TRENDS). In these two cases the path coefficients for marketing competencies are not significant. However, this is not to suggest that market knowledge does not play any role whatsoever in technologically more complex projects. It can be clearly seen that complementary competencies - as an integrator of both technological and marketing knowledge - are statistically significant in all of the models. This result is partially aligned with the findings of Lynn et al. (1996), which suggest that the use of commonly known market tools - among them concept testing, customer surveys, conjoint analysis, focus groups, and demographics segmentation- is limited when developing innovative products as they rely on users being able to articulate their needs. Furthermore, in the model accounting for the trend-setting strategy complementary competencies even outperform the technological competencies what is the opposite of the radical innovation model.

Figure 6: Baseline model of innovative performance and path coefficients

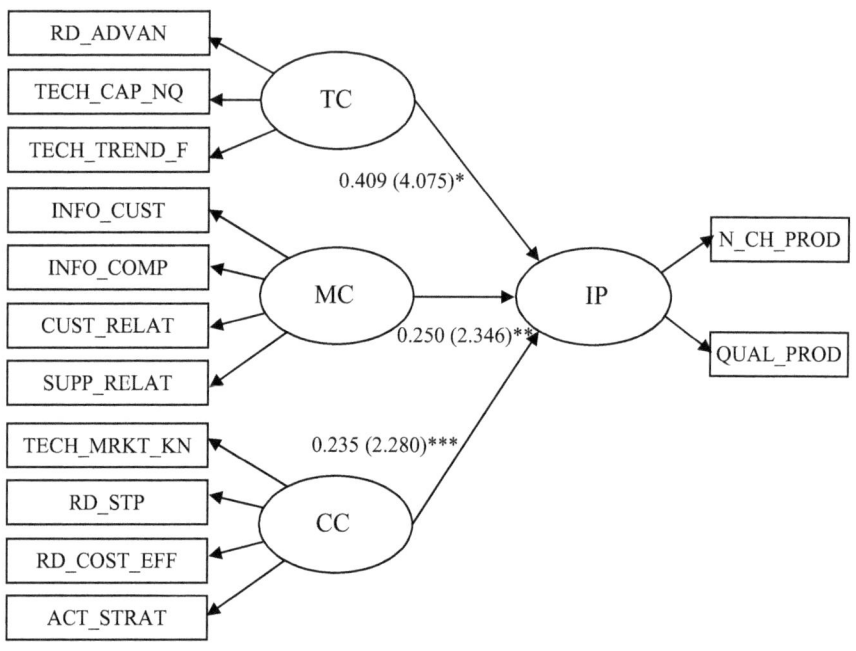

Note: T-values are stated in parentheses.
 * Significant at level P<0.001.
 ** Significant at level P<0.01.
 *** Significant at level P<0.05.

Table 16: Comparison of path coefficients of the constructs for the three models

Path	Baseline model N_CH_PROD	Incremental innovation model TIME_IMPR	Radical innovation model TIME_NEW	Model accounting for trend-setting TRENDS
TC→IP	0.409 (4.075) *	0.341 (2.843) **	0.363 (2.787) **	0.307 (2.537) **
MC→IP	0.250 (2.346) **	0.211 (1.537) ****	0.137 (1.192)	0.115 (0.813)
CC→IP	0.235 (2.280) ***	0.301 (2.423) **	0.352 (2.769) **	0.381 (2.582) **
R^2	0.63	0.57	0.58	0.52

Note: T-values are stated in parentheses.
* / ** / *** / **** P<0.001, P<0.01, P<0.05 and P<0.1, respectively.

Table 17 lists, for each of the four models, the weights of specific indicators with respect to their corresponding latent variables, thus making it possible to take a more detailed look at the competencies. These weights explain the link between the manifest variables and their latent counterparts.

Table 17: Weights of manifest variables for the four models

Indicator	Baseline model N_CH_PROD	Incremental innovation model TIME_IMPR	Radical innovation model TIME_NEW	Model accounting for trend-setting TRENDS
RD_ADVAN	0.3257	0.3371	0.3551	0.3237
TECH_CAP_NQ	0.4074	0.4234	0.3984	0.4181
TECH_TREND_F	0.3874	0.3608	0.3688	0.3786
INFO_CUST	0.4289	0.4007	0.4020	0.4012
INFO_COMP	0.2203	0.2752	0.2960	0.2662
CUST_RELAT	0.3490	0.3368	0.3262	0.3396
SUPP_RELAT	0.2897	0.2893	0.2808	0.2936
TECH_MRKT_KN	0.3390	0.3507	0.3306	0.3334
RD_STP	0.3524	0.3405	0.3772	0.3284
RD_COST_EFF	0.2852	0.3067	0.2962	0.2728
ACT_STRAT	0.3571	0.3391	0.3323	0.3948

Based on the obtained weights, the latent variables for the baseline model could also be written as follows:

$$TC = 0.3257 * RD\_ADVAN + 0.4074 * TECH\_CAP\_NQ + 0.3874 * TECH\_TREND\_F \quad (16)$$

$$MC = 0.3490 * CUST\_RELAT + 0.2203 * INFO\_COMP + 0.4289 * INFO\_CUST + \\ + 0.2897\ SUPP\_RELAT \quad (17)$$

$$CC = 0.3571 * ACT\_STRAT + 0.2852 * RD\_COST\_EFF + 0.3524 * RD\_STP + \\ + 0.3390 * TECH\_MRKT\_KN \quad (18)$$

$$IP = 0.6396 * NO\_CH\_PROD + 0.6015 * QUAL\_PROD \quad (19)$$

In the baseline model of innovative performance, the indicator TECH_CAP_NQ has the largest influence on the construct of technological competencies. The access to a number of different quality technological capabilities has a beneficial effect on new-product variety. It is interesting to note that the advancement of R&D (RD_ADVAN) comes last, even after the indicator of predicting technological trends (TECH_TREND_F).

Firms wishing to accelerate new-product development should according to Zahra & Ellor (1993) combine both radical and incremental innovation capabilities, which makes advanced R&D capabilities an indispensable element of the process. However, the performance of a higher novelty development process is, in turn, both more uncertain and more risky, although such projects tend to yield high returns if successfully commercialized (Mansfield & Wagner, 1975). The causes of this uncertainty are technically unfeasible project goals and insufficient market demand. Therefore, R&D activities may not necessarily be as effective when measured in terms of innovative performance.

While the weight of the variable TECH_CAP_NQ remains the highest of technological competencies indicators in all four models, it has the lowest value within the radical innovation model. The indicator that simultaneously, and conversely, appears to gain the most weight in this same model is RD_ADVAN. Technological novelty and superiority are thus prerequisites for the development of completely new products that are new generation products.

The importance of customer orientation is confirmed through marketing competencies. INFO_CUST and CUST_RELAT are the two key marketing competencies throughout the models. In the model of incremental innovation some of the weight of INFO_CUST is lost relative to INFO_COMP. As incremental innovations tend to be closely connected to imitation (Schewe, 1996), information regarding the activities of competitors can be an important guideline aiding in the formulation of R&D strategy and generation of new products. The relative importance of INFO_COMP also increases in the last two models; however, they allow only limited conclusions to be drawn since the relation between marketing competencies and innovative performance is not statistically significant.

In the group of complementary competencies for the baseline model, it is the indicators ACT_STRAT and RD_STP that stand out. It can be concluded that innovation strategy not only has to be a clearly stated strategy of a firm but also well defined. RD_STP can be viewed as an extension of the technological competencies indicator TECH_CAP_NQ by including the external environment of the firm as already explained in the segmentation results. While developing new technological capabilities in-house can prove to be very costly both financially and time wise, cooperation in R&D with external partners offers a viable alternative, especially to those companies that could otherwise not afford R&D at all (Hagedoorn, 2002). Involving suppliers in product design both early and extensively can serve to reduce the complexity of the design project, resulting in faster and more productive R&D processes (Gupta & Wileman, 1990). Customer involvement also notably improves the effectiveness of the product concept (Zirger & Maidique, 1990).

The cost efficiency of R&D (RD_COST_EFF) contributes the least of all complementary competencies. Although integration of technological and marketing knowledge can positively influence the efficiency of the development processes, it is also possible that due to the complexity arising from such coordination the processes become lengthier and more costly.

The model of incremental innovation differentiates itself decisively from the baseline model in terms of the variable TECH_MRKT_KN being of primary importance. This finding very much represents what the essence of incremental innovations is; namely, addressing different market needs by producing a variety of products within the same product family. Since incremental innovations are less costly and technologically demanding, it is also to be expected that RD_COST_EFF gains some importance relative to other indicators.

RD_STP is the indicator with the highest weight among complementary competencies within the third model – the radical innovation model. As the knowledge base requirements for the development of the most advanced products grows, strategic partnerships appear to be of increasing importance in facilitating the R&D activities. Access to technological capabilities may prove to be particularly problematic in a small economy, such as that of Slovenia. The companies are relatively small compared to their international counterparts and have smaller funds available for the financing of their R&D. Strategic partnerships are a way to gain access to additional capabilities through much smaller investments. The result is in line with the finding of Tidd and Bodley (2002), who confirmed, in the cases of both customer and user, that partnerships are more effective for high-novelty projects than for low-novelty ones.

It is interesting to note that the variable ACT_STRAT is the main driver of complementary competencies for the trend-setting model. It implies that clear strategic orientation is key when pursuing this position in the industry. The next most important variable in this model is TECH_MRKT_KN, stressing again the importance of the integration of both technological and marketing capabilities. Understanding the market nevertheless appears to be of vital importance. The smallest relative weight is assigned to RD_COST_EFF. The strategy of being an industry leader proves to be incompatible with building a competency based on cost efficiency in R&D.

Through the confirmation of the validity of the operational model of innovative performance using SEM and PLS, the validity of all constructs included in the model was confirmed for all four models of innovative performance; more specifically, the baseline model, the model of incremental innovation, the model of radical innovation, and the model accounting for trend-setting. With the exception of the models of radical innovation and trend-setting, where the relationship between marketing competencies and innovative performance were not statistically significant, all models exhibited statistically significant and positive links between competencies and innovative performance. This confirms the first and main hypothesis that innovative performance is affected by the three groups of competencies. The absence of a link between the marketing competencies and innovative performance in the technologically most demanding models (radical innovation and trend-setting) is not sufficient to render marketing competencies obsolete, as they are also - albeit indirectly - strongly present in complementary competencies. These two models are also the most representative of radical innovation, where the expected importance of technological competencies is, unlike for incremental innovations, high. This is in line with Hypothesis 5.

Hypothesis 6 is only partially supported. The models of radical innovation and trend-setting show the strongest link between technological competencies and the availability of different technological capabilities, followed by the competence of predicting technological trends. Nevertheless, the model of radical innovation does place the highest proportional weight on advancement of R&D. In this model, competence in strategic technological partnerships is also the key complementary competence with respect to the facilitation of highly demanding new-product development activities. Participating in strategic technological partnerships is also the second most important complementary competence in the baseline model - where it ranks behind alignment with strategy - and in the model of incremental innovation, in which it is positioned after the transfer of technological and marketing knowledge. Competence in strategic technological partnerships comes only third in the model accounting for trend-setting, behind the both highly recognized competencies of alignment with business strategy and transfer of technological and marketing knowledge. Hypothesis 7 is thus supported but only limitedly conclusive in the case of the trend-setting model, wherein the competence regarding innovative strategy is of greater importance.

6.8 Extensions of the baseline model

6.8.1 Complementary competencies as interaction between technological and marketing competencies

In this chapter, Hypothesis 9 is tested, whether complementary competencies can, in fact, be replaced by an interaction between technological and marketing competencies as proposed by Song et al. (2005) or if they should be considered as an independent group of competencies (Wang et al., 2004). What is particularly important about this distinction is the implication as to how these competencies should be treated within a firm. If complementary competencies are not a unique set of competencies, then there is no need to foster their development and, therefore, companies should focus on technological and marketing competencies alone as a means of managing innovative performance.

Applying SEM, internal consistency reliability, convergent validity and discriminant validity were all confirmed for this restricted version of the baseline model of innovative performance (Figure 7). Results are given in Table 18 and Table 19.

Figure 7: Restricted baseline model – complementary competencies excluded – as tested for validity

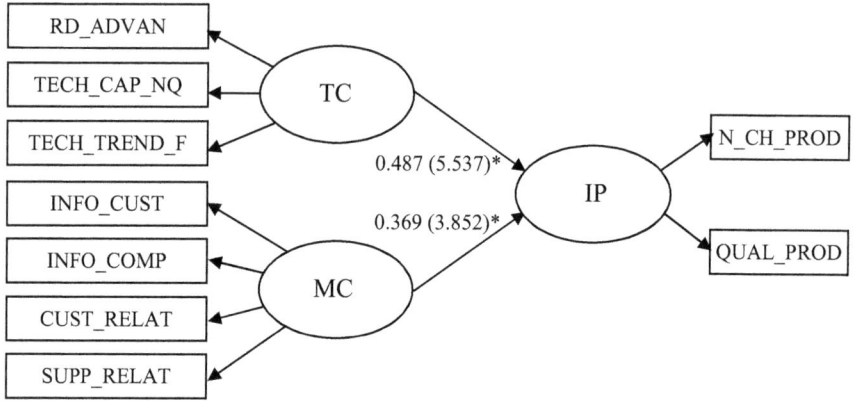

Note: T-values are stated in parentheses.
* Significant at level P<0.001.

Table 18: Composite reliability, correlation matrix and the square roots of AVE

	Composite Reliability	TC	MC	IP
TC	0.9198	*0.8903*		
MC	0.8438	0.6545	*0.7614*	
IP	0.7865	0.7287	0.688	*0.8052*

Note: The square roots of AVE are in the diagonal in italics. Below the diagonal are correlation coefficients.

Table 19: Cross loadings between manifest and latent variables

Indicators	TC	MC	IP
RD_ADVAN	0.8485 (20.592)		
TECH_CAP_NQ	0.9073 (43.024)		0.7018
TECH_TREND_F	0.9139 (43.946)		0.6691
INFO_CUST		0.8457 (27.290)	0.6741
INFO_COMP		0.6150 (5.812)	
CUST_RELAT		0.8740 (30.112)	
SUPP_RELAT		0.6794 (8.427)	
N_CH_PROD	0.6427		0.8300 (19.724)
QUAL_PROD			0.7796 (11.150)

Note: Mean of absolute values of cross loadings is 0.6379.
T-values are stated in parentheses for those indicators that belong to a designated latent variable in the model.
All significant at P<0.001.

In the second stage, the influence of the interaction term of technological and marketing competencies on the innovative performance was tested. The interaction term is constructed using technological competencies as a predictor variable and marketing competencies as a moderator variable (Figure 8). The interaction term was standardized, as proposed by Chin et al. (2003, p. 198-199), to help avoid computational errors by lowering the correlation between the product indicator and their individual components. The methodology applied follows procedures suggested by Tabachnick and Fidell (2007, p. 728-729).

Figure 8: Inclusion of interaction term of technological and marketing competencies

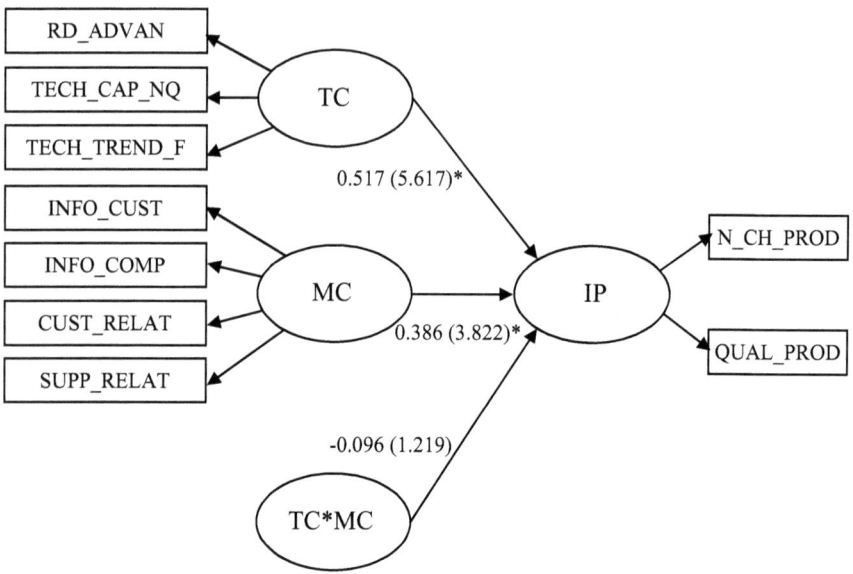

Note: T-values are stated in parentheses.
        * Significant at level P<0.001.

While a positive main effect of technological (β=0.487; significant at P<0.001) and marketing competencies (β=0.369; significant at P<0.001) on innovative performance can again be confirmed, the interaction term used as a proxy for complementary competencies does not have a statistically significant effect. By including the interaction term, the value of R^2 increased from 0.609 to 0.616 (Table 20). However, this increase is not significant $[F(1,62)=1.13] < [F_{critical} = 3.99]$.[1] The effect size is negligible (Cohen & Cohen, 1983, p. 155-158).[2]

---

1   $\Delta F = (\Delta R^2 (C-p^*)) / (q (1-R^2 current))$, where C is the number of observations, p* the number of coefficients in the model, and q the number of added independent variables.
2   Effect size $f^2 = (R^2 \text{interaction model} - R^2 \text{main effects}) / R^2 \text{interaction model}$.

Table 20: PLS path analysis results: effect of the interaction term of technological and marketing competencies on innovative performance

Exogenous variables	Stage I	Stage II
TC	0.487 (5.035)*	0.517 (5.617)*
MC	0.369 (3.543)*	0.386 (3.822)*
TC x MC		-0.096 (1.219)
R^2	0.609	0.616
F (1, 62)		1.13
Effect size f^2		0.01

Note: * Significant at level P<0.001.

This result is conclusive with the validity test of the baseline model, where complementary competencies have already been confirmed as a valid construct. In addition, the test of the interaction term provides further support for Hypotheses 8 and 9.

6.8.2 Extension of the baseline model for business performance

In order to analyze how innovative performance contributes to the business performance of a firm, the whole operational model was tested as presented in Figure 9, by including the general construct of innovative performance from the baseline model, as measured by NO_CII_PROD and PROD_QUAL. The proposed model was assessed for the weighted sample of 50 firms, since business performance measures were collected for firms as a whole. Responses regarding the competencies in innovative-performance measures of those firms that reported multiple product lines were weighted, and the weights assigned corresponded to the share of a specific product line in total sales.

The validity of the model was checked in the same way as previously described. Internal consistency reliability was confirmed. Values of composite reliability for all constructs exceed the threshold of 0.70, the minimum value being 0.7912 (Table 21).

In Table 22 only cross loadings with values larger than the mean of the absolute values, 0.5113, are shown. The minimum value of cross loadings for the proposed indicators in the observed model is 0.6073, above the 0.60 threshold. All latent variables are again well correlated with their own indicators. AVE for each construct is higher than 0.50 (see the square roots of AVE in Table 21). Furthermore, they are all higher than the correlation coefficients below them. This confirms discriminant validity. Standard errors were computed on the basis of 500 bootstrapping runs and 50 cases.

Table 21: Composite reliability, correlation matrix and the square roots of AVE

	Composite reliability	TC	MC	CC	IP	BP
TC	0.9175	*0.8875*				
MC	0.8497	0.6138	*0.7677*			
CC	0.7998	0.6377	0.6776	*0.7080*		
IP	0.7912	0.6988	0.6595	0.6628	*0.8094*	
BP	0.7916	0.2628	0.5506	0.4025	0.4784	*0.8139*

Note: The square roots of AVE are in the diagonal in italics. Below the diagonal are correlation coefficients.

As shown in Figure 9, technological competencies have the largest influence on innovative performance (β=0.386, significant at P<0.01), followed by marketing and complementary competencies (the values of whose correlations are β=0.259 and β=0.241 respectively; both significant at P<0.05). The path coefficients are aligned with the findings of the partial baseline model of innovative performance already explained (Table 16). The model also confirms the influence of innovative performance on business performance with the path coefficient being 0.478 (significant at P<0.001). The value of R^2 for innovative performance is 60.0% and for business performance 23%.

To conclude, there exists a positive link between innovative performance and firm performance which is statistically significant. Product variety and technical performance (quality) inherent in innovative performance contribute to a firm's bottom line. Hypothesis 10 is thus supported.

Table 22: Cross loadings between indicators and latent variables

Indicators	TC	MC	CC	IP	BP
RD_ADVAN	**0.8493** (**15.181**)		0.5289	0.5455	
TECH_CAP_NQ	**0.9009** (**30.653**)	0.5706	0.5243	0.6575	
TECH_TREND_F	**0.9111** (**38.727**)	0.5831	0.6427	0.6487	
INFO_CUST	0.5500	**0.8478** (**25.985**)	0.5774	0.6450	0.5468
INFO_COMP		**0.6191** (**4.637**)			
CUST_RELAT	0.5468	**0.8504** (**18.842**)	0.5283		
SUPP_RELAT		**0.7295** (**8.058**)	0.5897		
TECH_MRKT_KN		0.5159	**0.7481** (**7.915**)		
RD_STP	0.6211		**0.7183** (**8.443**)		
RD_COST_EFF			**0.6073** (**4.152**)		
ACT_STRAT		0.6394	**0.7489** (**9.314**)		
N_CH_PROD	0.6681			**0.7678** (**5.972**)	
QUAL_PROD		0.5758	0.5806	**0.8490** (**13.826**)	0.5336
AVG_ROA_0406		0.5647			**0.9476** (**31.496**)
AVG_ROE_0406					**0.6534** (**4.501**)

Note: T-values are stated in parentheses for those indicators that belong to a designated latent variable in the model. All significant at P<0.001.

Figure 9: Operational model of innovative and business performance
with path coefficients between latent variables

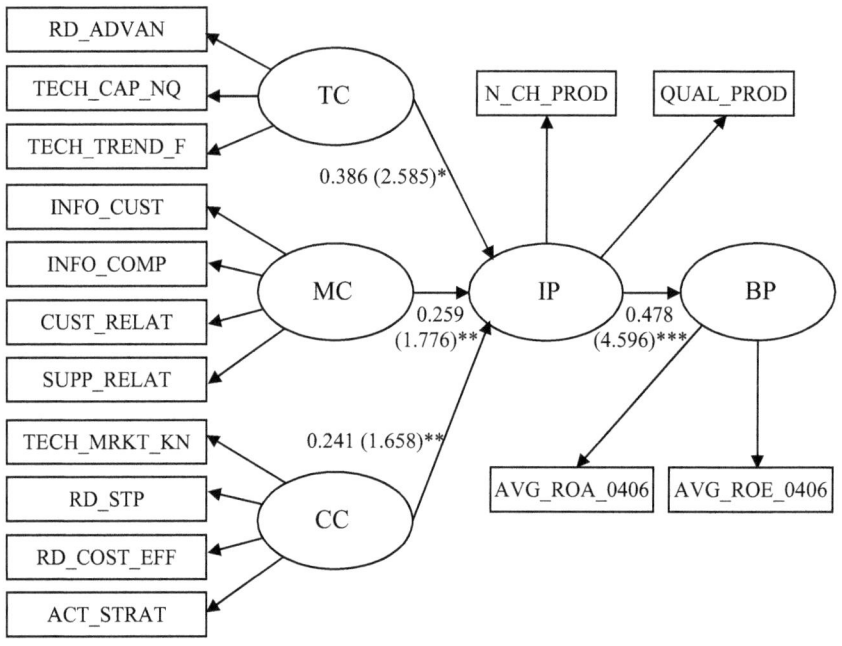

Note: T-values are stated in parentheses.
* Significant at level P<0.01.
** Significant at level P<0.05.
*** Significant at level P<0.001.

6.8.3 Sampling bias

Sampling bias occurs due to distortions in the collection of observations constituting a sample. The bias that could be of the most relevance herein is that of self-selection. Bigger companies could, on the one hand, be more interested in participating since they are more confident of their achievements as well as more interested in learning from the eventual results of this study.

A Probit model for survey participation was employed first. A participation dummy variable was regressed on variables that are considered to influence the decision of firms to participate in the survey. At the same time these data also have to be available for all non-responding firms. The analysis was thus performed on 328 non-responding firms and 50 responding firms. Drawing on data from the AJPES database, the following variables were included: natural logarithm of total sales, capital intensity and share of exports in total sales.

Total sales are one of the key indicators of a firm's size. Capital intensity was calculated as fixed assets spent per employee. It takes into account the bias that could be attributed to the differences in the nature of the industries, namely whether they are capital or labour intensive, with capital intensive industries achieving higher value added. Firms in these industries may have a clearer idea of the drivers of competitive advantage and may not be as interested in learning from such research results as will become available to them afterwards. The rationale for the inclusion of the last variable is that firms with a larger share of exports could be considered more competitive and thus more successful, and therefore more inclined to participate in the survey. That is why this variable was also included. Since variables number of employees and value of exports were not available after the year 2004, data from this last available year were used. The results of this regression are presented in Table 23. Other things being equal, firms with higher sales and lower capital intensity are more likely to participate in the survey.

Table 23: Probit model for survey participation

Dependent variable	Response dummy
Constant	-4.15 (0.000)*
Ln(Sales)	0.208 (0.004)*
Capital intensity	-0.00002 (0.07)**
Share of exports in total sales	0.425 (0.156)

Note: Pseudo R^2 = 7.4%. T-values are stated in parentheses.
* / ** Coefficient estimates are significantly different from zero at the 1% and 10% level, respectively.

The inverse Mill's ratio was further calculated as:

$$\lambda = \frac{\phi(Z)}{\Phi(Z)} \tag{20}$$

Where ϕ is a probability density function and Φ the cumulative density function of the standard normal distribution. Z are the fitted values from the Probit equation calculated as (Greene, 2003, p. 784):

$$Z = \beta_{10} + \sum_i \beta_{1i} x_{1i} \tag{21}$$

A Mill's ratio was calculated for each company as a whole, and thus the same Mill's ratio is applied to every product line from the same firm. A baseline model of innovative performance was tested on a sample of 65 product lines, by including Mill's ratio as an additional indicator with each construct of competencies (Figure 10). Mill's ratio was included both to check for and correct potential sampling bias in the sample.

While the internal consistency reliability can be confirmed, Mill's ratio as a manifest variable does not appear to adequately describe the respective latent variables. AVE is lower than 0.50 for the latent variable marketing competencies as well as for complementary competencies.

As the square root of AVE is not higher than correlations with other variables in the case of marketing and complementary competencies, the criterion of discriminant validity is not satisfied (Table 24). Furthermore, Mill's ratio as a variable did not prove to be statistically significant (Table 25).

By testing the weighted firm sample (N=50) for sampling bias a similar conclusion can be drawn. Internal consistency reliability can again be confirmed as the values of composite reliability for all constructs exceed the threshold of 0.70. AVE values for marketing and complementary competencies are again lower than the threshold value of 0.50 (square roots of AVE given in Table 26) violating discriminant validity. Mills' ratio is also not statistically significant as a manifest variable in the model. The path coefficients obtained were 0.426 (P<0.001) for technological competencies, 0.240 (P<0.05) for marketing competencies and 0.216 (P<0.1) for complementary competencies. The model featuring the addition of Mill's ratio correcting for sampling bias is again invalid and we can assume there is no evident sampling bias present (Table 27).

Figure 10: Baseline model of innovative performance with the inclusion of the Mill's ratio variable and corresponding path coefficients for the sample of 65 product lines

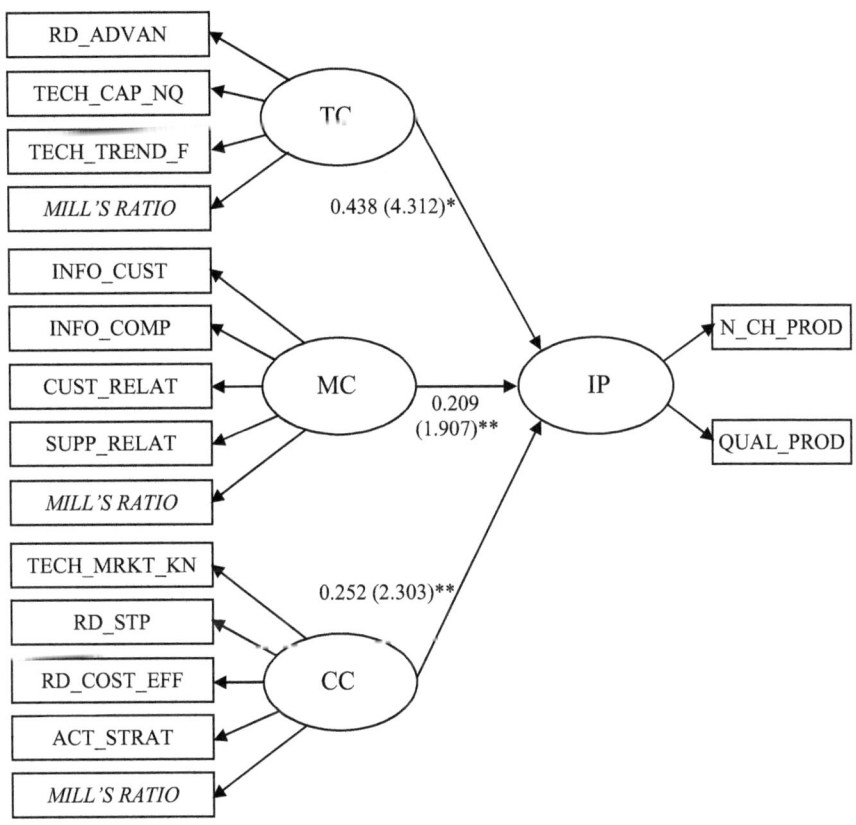

Note: T-values are stated in parentheses.
* Significant at level P<0.001.
** Significant at level P<0.05.

Table 24: Composite reliability, correlation matrix and the square roots of AVE (product lines)

	Composite reliability	TC	MC	CC	IP
TC	0.8210	*0.7712*			
MC	0.7828	0.6566	*0.6799*		
CC	0.7841	0.6604	0.7231	*0.6721*	
IP	0.7868	0.7418	0.6793	0.6929	*0.8054*

Note: The square roots of AVE are in the diagonal in italics. Below the diagonal are correlation coefficients.

Table 25: Cross loadings between manifest and latent variables (product lines)

Indicators	TC	MC	CC	IP
RD_ADVAN	**0.8471** **(17.722)**	0.5298	0.5698	0.5833
TECH_CAP_NQ	**0.9068** **(44.449)**	0.5954	0.5745	0.7118
TECH_TREND_F	**0.9141** **(35.010)**	0.6213	0.6168	0.6764
INFO_CUST	0.6331	**0.8413** **(24.864)**	0.6325	0.6739
INFO_COMP		**0.6245** **(5.794)**		
CUST_RELAT	0.5469	**0.8695** **(25.970)**	0.6167	0.5326
SUPP_RELAT		**0.6677** **(8.544)**	0.5088	
TECH_MRKT_KN		0.6036	**0.7911** **(11.097)**	0.5191
RD_STP	0.6583	0.5564	**0.7589** **(17.721)**	0.5472
RD_COST_EFF			**0.6277** **(5.130)**	
ACT_STRAT	0.5162	0.6809	**0.7928** **(12.536)**	0.5574
N_CH_PROD	0.6583	0.5460	0.5274	**0.8213** **(14.593)**
QUAL_PROD	0.5325	0.5489	0.5917	**0.7891** **(12.056)**
MILLS RATIO	0.0589 (0.299)	0.1092 (0.609)	0.1846 (1.094)	

Note: Mean of absolute values of cross loadings is 0.5066.
T-values are stated in parentheses for those indicators that belong to a designated latent variable in the model. All except Mill's ratio significant at P<0.001.

Table 26: Composite reliability, correlation matrix and the square roots of AVE (firm)

	Composite reliability	TC	MC	CC	IP
TC	0.8294	*0.7717*			
MC	0.8070	0.6177	*0.6948*		
CC	0.7628	0.6438	0.6777	*0.6425*	
IP	0.7925	0.7138	0.6502	0.6535	*0.8101*

Note: The square roots of AVE are in the diagonal in italics. Below the diagonal are correlation coefficients.

Table 27: Cross loadings between manifest and latent variables (firm)

Indicators	TC	MC	CC	IP
RD_ADVAN	**0.8482** **(18.062)**		0.5302	0.5717
TECH_CAP_NQ	**0.8995** **(32.517)**	0.5703	0.5264	0.6696
TECH_TREND_F	**0.9174** **(30.604)**	0.5857	0.6431	0.6567
INFO_CUST	0.5546	**0.8484** **(23.756)**	0.5778	0.6388
INFO_COMP		**0.6171** **(5.020)**		
CUST_RELAT	0.5477	**0.8512** **(20.778)**	0.5243	0.5065
SUPP_RELAT		**0.7276** **(8.351)**	0.5861	
TECH_MRKT_KN		0.5191	**0.7441** **(8.268)**	
RD_STP	0.6201		**0.7260** **(12.051)**	0.5086
RD_COST_EFF			**0.6173** **(4.519)**	
ACT_STRAT		0.6404	**0.7344** **(9.023)**	
N_CH_PROD	0.6659			**0.8266** **(13.526)**
QUAL_PROD		0.5730	0.5729	**0.7934** **(9.400)**
MILLS RATIO	0.1446 (0.619)	0.2438 (1.083)	0.2506 (1.136)	

Note: Mean of absolute values of cross loadings is 0.5018.
T-values are stated in parentheses for those indicators that belong to a designated latent variable in the model. All except Mill's ratio significant at P<0.001.

6.8.4 Moderating effects of environmental turbulence

In this part of the study the baseline model of competencies, innovative performance and business performance is expanded to demonstrate the possible moderating effects of environmental variables; namely, technological and market turbulences. The aim is to analyze whether the moderating effects have any direct impact on the business performance of a firm and if there is any interaction effect with innovative performance. Additional constructs are presented in Table 28.

Table 28: Two environmental effects as latent variables and their indicators

Manifest variable (MV)	MV label	Latent variable (LV)	LV label
New technologies have a high impact on business operations and competition and bring about big opportunities.	NEW_TECH_OP	Technological turbulence	TT
Technology in our industry is changing rapidly.	TECH_CH		
It is almost impossible to predict accurately the rapidly changing tastes and demands of consumers.	CH_DEMAND	Market turbulence	MT
The level of market uncertainty is extremely high.	MKT_UNC		

The indicators of both technological and market turbulence that were included in the questionnaire were based on existing literature (Wang et al., 2004; Calantone et al., 2003). However, bearing in mind the cut-off for factor loadings, 0.60 (Hatcher, 1994), only 2 indicators per each latent variable of environmental turbulence made it into the final model. For technological turbulence (TT), these were business potential of new technologies (NEW_TECH_OP) and the speed of change in the industry's technology (TECH_CH). The two variables not included in the model were: the predictability of technological changes in the next 2 to 3 years, and smaller innovations being the driver of technological advances. With respect to market turbulence, the two indicators included were: the predictability of changes in customer demand (CH_DEMAND), and the level of market uncertainty (MKT_UNC). Excluded were the variables referring to the predictability of major competitors' activities and the intensity of competition in the industry.

The measurement model (Figure 11) was again tested for internal consistency reliability, convergent validity and discriminant validity, all of which were confirmed (Table 29 and Table 30).

Figure 11: The model with technological and market turbulences as tested for validity

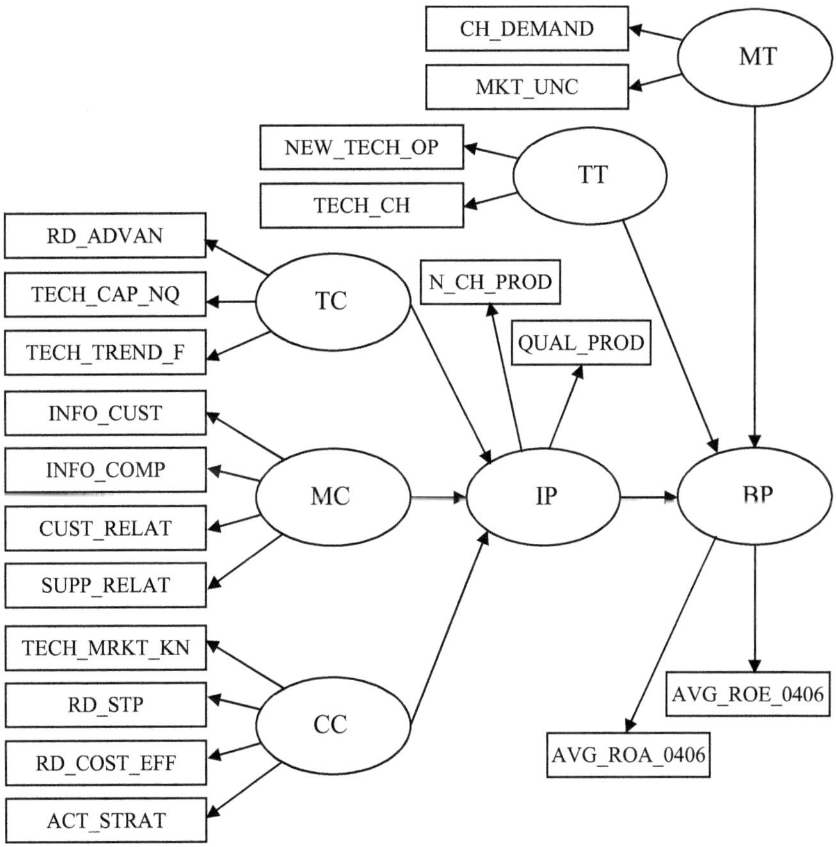

Table 29: Composite reliability, correlation matrix and the square roots of AVE

	Composite reliability	TC	MC	CC	IP	BP	TT	MT
TC	0.9175	*0.8876*						
MC	0.8497	0.6138	*0.7677*					
CC	0.7998	0.6386	0.6758	*0.7080*				
IP	0.7923	0.7055	0.658	0.6614	*0.8101*			
BP	0.9794	0.3065	0.5591	0.3793	0.4828	*0.9796*		
TT	0.8852	0.2224	0.0456	0.0625	0.2793	0.0848	*0.8912*	
MT	0.8653	-0.0403	-0.1659	-0.0501	-0.0663	-0.3267	-0.0937	*0.8735*

Note: The square roots of AVE are in the diagonal in italics. Below the diagonal are correlation coefficients.

Table 30: Cross loadings between manifest and latent variables

Indicators	TC	MC	CC	IP	BP	TT	MT
RD_ADVAN	**0.8504** **(18.387)**	0.4719	0.5314	0.5556			
TECH_CAP_NQ	**0.9005** **(32.702)**	0.5708	0.5253	0.6626			
TECH_TREND_F	**0.9106** **(39.679)**	0.5831	0.6421	0.6523			
INFO_CUST	0.5501	**0.8477** **(25.318)**	0.5764	0.6435	0.5993		
INFO_COMP	0.4897	**0.6183** **(4.971)**					
CUST_RELAT	0.5465	**0.8515** **(18.211)**	0.5268	0.5012			
SUPP_RELAT		**0.7291** **(8.468)**	0.5884	0.4791			
TECH_MRKT_KN		0.5159	**0.7461** **(8.290)**				
RD_STP	0.6212	0.4992	**0.7229** **(10.310)**	0.4984			
RD_COST_EFF			**0.6106** **(4.508)**				
ACT_STRAT		0.6392	**0.7434** **(9.313)**	0.499			
NO_CH_PROD	0.6682	0.4875	0.4897	**0.79** **(7.378)**			
QUAL_PROD	0.4843	0.5753	0.5785	**0.8297** **(13.553)**	0.484		
AVG_ROA_0406		0.5431		0.4788	**0.9777** **(107.076)**		
AVG_ROE_0406		0.5519		0.4677	**0.9815** **(129.984)**		
NEW_TECH_OP						**0.8696** **(3.258)**	
TECH_CH						**0.9123** **(3.362)**	
CH_DEMAND							**0.9173** **(14.458)**
MKT_UNC							**0.8275** **(5.311)**

Note: Mean of absolute values of cross loadings is 0.4585.
T-values are stated in parentheses for those indicators that belong to a designated latent variable in the model. All significant at P<0.001.

After confirming the validity of constructs, a separate test was carried out, namely regarding the influence of technological and market turbulence (moderator variables) for exogenous latent variable business performance (predictor variable) with innovative performance as the endogenous latent variable. Interaction term of innovative performance and the selected latent variable of environmental effects – technological turbulence and market turbulence, respectively – was included in the second stage (Figure 12 and Figure 13). The interaction term was standardized, in line with Chin et al. (2003, p. 198-199). This methodology was applied to the analysis of environmental effects by Wang et al. (2004; p. 268) and follows procedures suggested by Tabachnick and Fidell (2007, p. 728-729).

Figure 12: The moderating effect of technological turbulence on innovative performance

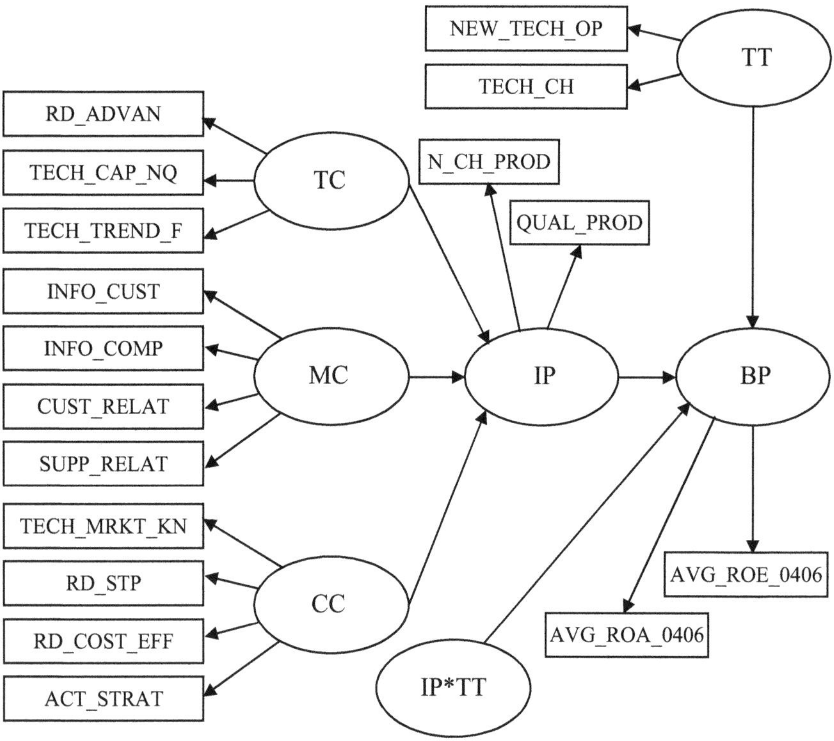

Figure 13: Moderating effect of market turbulence on innovative performance

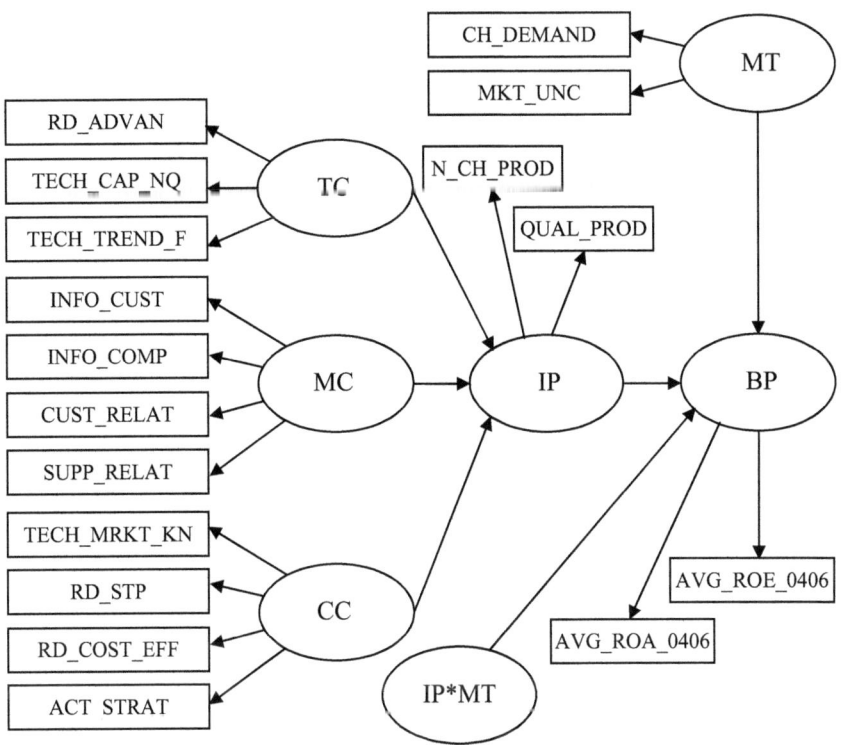

When technological turbulence (TT) was taken into account, the model's R^2 value in stage I was 0.236. With the inclusion of market turbulence (MT) it was 0.320 (Table 31). Innovative performance was significant at β=0.498 (P<0.001) and β=0.463 (P<0.001) respectively. It is therefore possible to conclude that there is significant support for there being a positive main effect of innovative performance on business performance. By adding the interaction term to the models, values of R^2 in second stage increased to 0.317 and 0.397 respectively. In both cases the increase of R^2 is significant at P<0.05 for technological and market turbulence ($F_{critical}$ = 4.06), both of which moderate the relationship between innovative performance and business performance.

Table 31: PLS path analysis results: the moderating effects of technological and market turbulence on the relationship between innovative performance and business performance

Variables	Technological turbulence (TT)		Market turbulence (MT)	
	Stage I	Stage II	Stage I	Stage II
IP	0.498 (4.390)*	0.434 (4.229)*	0.463 (4.729)*	0.450 (3.696)*
MT			-0.296 (2.782)**	-0.332 (2.658)**
TT	-0.052 (0.363)	-0.013 (0.089)		
IP x MT				-0.280 (1.001)
IP x TT		0.291 (1.609)***		
R^2	0.236	0.317	0.320	0.397
F (1, 44)		5.22		5.62
Effect size f^2		0.25		0.19

Note: * Significant at level P<0.001.
** Significant at level P<0.01.
*** Significant at level P<0.1.

From the results of the second stage of the analysis it can be observed that technological turbulence does have a positive moderating effect via innovative performance, though it is of marginal significance. The effect size $f^2 = 0.25$ is, according to Cohen and Cohen (1983, p. 155-158), medium. The higher the technological turbulence is, the greater the positive effect (positive value of the path coefficient of the interaction term) of innovative performance on firm performance will be.

With respect to the effect of market turbulence, it was discovered that it does indeed have a negative direct effect on the business performance of a firm. Following the inclusion of the interaction term, this direct effect is still present. The increase of R^2 attributable to marketing turbulence is statistically significant; however, the path coefficient between innovative performance and business performance is not. Nor is the moderating effect, although there appears to be a small effect according to the effect size $f^2 > 0.1$. Thus, market turbulence negatively affects business performance. This finding is not surprising; however, it can also be noted that by enhancing marketing efforts regarding the acquisition of market information, firms can, to a certain extent, decrease this uncertainty.

The small impact of market turbulence observed is in line with the findings of Hult et al. (2004). The authors confirmed that innovativeness is a determinant of business performance regardless of the market turbulence to which the firm is exposed.

Due to its composition, environmental turbulence can also, in a way, be viewed as a proxy of industry as a variable in the analysis. The small effect of these factors is in alignment with the findings of Deshpandé and Farley (2004, p. 14), who studied organizational culture, market orientation and firm performance in 12 different countries. They divided firms into seven groups according to their industries, namely: financial services, other services, consumer durables, consumer nondurables, and industrial products subdivided into: capital goods, equipment and supplies. The inclusion of industry as a covariate had no significant effect.

The support lent to Hypothesis 11 is limited since the moderating effects of technological and marketing turbulence observed are marginal. These findings are in accordance with Hypotheses 12 and 13 as the impact of technological turbulence is positive while that of marketing turbulence is negative.

7 Conclusion

Successful product innovation and the ability of firms to continuously improve their innovation processes are rapidly becoming key ingredients of competitive advantage and long term growth for companies in both the manufacturing and service sectors (Chapman & Hyland, 2004, p. 553). The competence-based view offers an insight into the drivers behind the competitive advantage.

Segmentation performed by means of hierarchical clustering revealed that the most innovative companies – technology leaders – demonstrate the presence of all competencies to a high degree of development. Two segments of followers that rely predominantly on imitation in their innovation strategy – technology followers with weak competencies, and technology followers with strong competencies – were also identified. The marketing competence with respect to which no differences among the three segments are exhibited is that of access to real-time information on competitors. This implies that regardless of innovative performance, companies are aware of the importance of understanding the prevailing dynamics in the industry. Furthermore, access to real-time information on competitors no longer appears to be a potential source of competitive advantage.

The distinctive core competencies of technology leaders are clearly technological competencies, whereas strong followers build their competitiveness on marketing and complementary competencies. Cooperating in strategic technological partnerships, and thus broadening the scope of technological capabilities at one's disposal, is an important, distinctive complementary competence for technology leaders. They are innovators and perform very favourably regarding new product development lead times and are notable contributors to industry trends.

The descriptive analysis for aggregate data already implied that Slovenian firms are operating in dynamic environments and are, generally, taking an active part. That is to say, firms witnessed, on average, short product life-cycles as well as intense competition through both R&D and design. On average, firms generate the vast majority of their sales through products branded as their own. Not only is this encouraging as it is important for the firms to enhance their brands' recognition, but also because own-brands facilitate higher price mark-ups. The motivation behind innovations is more defensive than offensive, hence the predominant incremental innovations. Enhanced company image and improved appearance rank as top motivations, a fact which points to the role of design.

The study further set out to establish which competencies firms develop and employ when pursuing different innovation strategies. The findings suggest that companies attempting to improve their innovative performance should focus first and foremost on technological competencies. The availability of various high-quality technological capabilities was recognized as the most decisive dimension contributing to new-product variety and quality. That said, marketing and complementary competencies should by no means be overlooked. From the viewpoint of marketing competencies, the greatest share of attention should be focused on customer-related competencies that guide the new-product development process towards best addressing customer needs. Among complementary competencies, companies should take particular care in ensuring they have a clear strategic direction. Strategic technological partnerships represent another key factor facilitating the expansion of a firm's access to different technological capabilities. It is also worth noting that a clear and well-defined strategy can help firms recognize their core competencies so as to be in a position to make a well-informed strategic management decision regarding the outsourcing of non-core competencies (Hafeez et al., 2007).

Studies on the state of R&D in Slovenian firms show that the economy falls into the category of a technology follower country (European Innovation Scoreboard, 2008, p. 7). As a member of the European Union, Slovenia is actively involved within The Lisbon strategy, an action and development plan aimed at increasing the competitiveness of the EU countries. On the basis of the research presented in the thesis, it is possible to draw several conclusions that support strategies proposed by the Agenda. For instance, although it may be costly and time consuming for technology follower countries to acquire technological competencies, marketing and complementary competencies can successfully facilitate the process of catching up via incremental innovation. Firms can thus opt for imitation as a strategy for developing technological capabilities, and thereby bridge the gap to a certain extent. This finding shares some common ground with the results of Armbruster et al. (2005) who observed in the case of German firms that they seem to be among the leaders in technical process innovations, whereas firms in the new member states of the European Union are utilizing innovative organizational forms.

Furthermore, novel technologies require advanced R&D. Entering strategic technological partnerships proves to be almost an imperative in achieving this by enabling access to additional technological and marketing capabilities. Moreover, firms directing trends within their industries and acting as market leaders build their competitive advantage first and foremost on complementary competencies, followed by technological competencies. Market leaders complement and support their technological competencies in virtue of having a solid strategy, successfully integrating technological and marketing knowledge, and by expanding their access to capabilities through strategic technological partnerships.

Environmental effects, namely technological and market turbulence, have little impact on how innovative performance affects firm's business performance. This could mean that firms have an acute awareness and understanding of their environments. Furthermore, it could also imply that the perception of the conditions in their respective industries is uniform among competitors, thus transcending specific markets.

7.1 Contribution to theory and practice

The main contribution made to the theory of competitive advantage and innovation is the validated model of technological, marketing and complementary competencies being linked to innovative performance and, furthermore, to business performance. From the theoretical point of view this is the first attempt to link all three concepts within the parameters of the same model. As such, it not only supports the positive link between innovation and business performance which is the objective of innovative activity, but also sheds light on the underlying competencies or, in other words, the competitive strengths of innovation within a firm. Technological, marketing and complementary competencies were chosen in order to best account for the key knowledge, skills and capabilities that are central to new-product development.

From the practical point of view this approach of measuring firm competencies can be useful because of the many opportunities it provides for data interpretation. By enabling cross-industry comparisons, country-level data can be analyzed. In this manner an insight into the dynamics of the economy is obtained. National policies often focus on select industries that are thought to have the greatest potential. The competencies and innovative performance approach clearly shows that companies' potential is not necessarily contingent on industry, but more likely on the competencies they are able to develop.

On the firm level, an aggregate analysis of competencies can provide firms with guidelines for their innovation strategy, as they can use the findings to work out and define their own innovation strategy. At that stage they can refer to the findings presented and identify the competencies they need to further develop. In order to understand which concrete actions are behind these competencies, they can make use of the approaches presented for breaking down competencies into industry- and firm-specific capabilities. This will help them identify concrete steps that need to be taken. The measures of competencies proposed, are of course, not limited to aggregate analysis on the country level, which is why firms can also use them independently for positioning themselves within a more limited context of their interest, for instance specific markets.

Development of the model alone also required a synthesis of literature which proved to be rather ambiguous in terms of the definitions in use. Definitions were streamlined and a corresponding set of measures was developed. The objective was to devise a set of simple straightforward measures that best encompass competencies and at the same time facilitate easy comparison among firms.

Although the small sample size could be considered a limitation of the study, all attempts were made to ensure its reliability through an elaborate survey design and questionnaire structure. The sample was also tested for any sampling bias. A further way to improve the reliability of the results would be to increase the sample size and potentially include more respondents from a single firm and weigh their responses. The measures used are also subject to further improvements and adjustments through continued research.

Competencies undoubtedly offer an insight into a firm's strategy for creating competitive advantage. However, it is important to keep in mind that sustainable competitive advantage is not a final destination a company can and should reach, but rather an ongoing, dynamic journey (Chaharbaghi & Lynch, 1999, p. 45). Therefore, companies need to constantly possess an understanding of how their competencies are positioned relative to their competitors and, furthermore, commit themselves to constantly enhancing them, especially those identified as core competencies. The core of a strategy for the creation of competitive advantage has to be twofold; namely striving to improve oneself in order to remain competitive while being unique so as to get ahead of the competition (Porter, 1999). This line of thinking is inherent in the concept of core competencies.

The presented research is based on firm data from the pre-recession period what may at first glance seem to have put concepts of innovation and firm competencies into a different strategic perspective, especially in terms of how their roles and relevance have changed. However, despite the new circumstance, there is still a common consensus among practitioners and researchers that innovation remains to be the one prerequisite for long term growth and is the expected driver of economic recovery (Filippetti & Archibugi, 2011; OECD, 2010). It has the potential to lead the firms out from the downward facing spirals of economic slowdown and the necessary strategic adaptability. Their core businesses brought them to where they were before 2008 and in the past 3 years the consumers' needs have not changed as much as their capacity to afford to satisfy those needs. Therefore, changed market characteristics may yield some of the old projects obsolete but there is no reason to believe that the underlying rules have changed altogether. As McKinsey&Company (2010) conclude from their 2010 Global Survey results on innovation and commercialization, the challenges companies face today have in fact not changed from before the crisis. They remain to be finding the right talent, encouraging collaboration and risk taking and organizing the innovation process from beginning to end.

Different authors propose a variety of approaches to innovation in the current environment but only time will reveal the most successful ones. Reeves and Deimler (2009) list behavioral strategies, such as adjusting the customer offering, and exploring new pricing offers, finding new markets and exiting old ones, as well as seizing M&A opportunities for acquiring assets. Among their social strategies are partnering with suppliers or customers, partnering with competitors, and shaping the regulatory environment. They further advocate reproductive strategies, for example rapid prototyping and incubation, new business models and resilience or adaptability. The latter emphasizes among others good monitoring of trends and forecasting, scenario planning and ability to adapt accordingly, and resources to foster staying power. Chesbrough and Garman (2009) place key importance on moving innovation from the inside out and letting nonstrategic initiatives to be developed externally. The proposal for improvements according to McKinsey&Company (2010) includes formalizing the processes for setting priorities and commercializing products as well as integrating innovation into firms' strategic-planning efforts. With regard to the concept of competencies, it is apparent that the above proposals call not only for strong technological and marketing competencies, but also strong complementary competencies, especially strategy wise.

References

Afuah, A. (2002). Mapping technological capabilities into product markets and competitive advantage: The case of cholesterol drugs. *Strategic Management Journal*, 23, 171-179.

Aho, E., Cornu, J., Georghio, L. & Subirá, A. (2006). *Creating an innovative Europe*. Luxembourg: Office for Official Publications of the European Communities. Retrieved November 18, 2008, from http://www.eua.be/Libraries/Page_files/aho_report.sflb.ashx

Ahuja, G. & Katila, R. (2001). Technological acquisitions and the innovation performance of acquiring firms: A longitudinal study. *Strategic Management Journal*, 22, 197-220.

AJPES (2007). *Firm database*. Ljubljana: Agency of the Republic of Slovenia for Public Legal Records and Related Services.

Akao, Y. (2004). *Quality Function Deployment: Integrating Customer Requirements Into Product Design*. Cambridge, Massachusetts: Productivity Press.

Alajääsko, P. (2008). Main features of the EU-27 services sector. *Eurostat – Statistics in Focus*, 78, 1-8.

Ali, A., Kalwani, M. U. & Kovenock, D. (1993). Selecting product development projects: Pioneering versus incremental innovation strategies. *Management Science*, 39 (3), 255-274.

Amit, R. H. & Schoemaker, P. J. H. (1993). Strategic assets and organizational rents. *Strategic Management Journal*, 14 (1), 33-46.

Anderson, J. C. & Gerbing, D. W. (1988). Structural equation modeling in practice: A review and recommended two-step approach. *Psychological Bulletin*, 103, 411-423.

Andrews, D. W. K. & Buchinsky, M. (2000). Three-step method for choosing the number of bootstrap repetitions. *Econometrica*, 68 (1), 23-51.

Andrews, K. R. (1998). *The concept of corporate strategy*. In Foss, N. J. (ed.), *Resources, firms, and strategies: A reader in the resource-based perspective*. New York: Oxford University Press, 52-59.

Ansoff, H. I. (1965). *Corporate Strategy*. New York: McGarw Hill.

Appiah-Adu, K. & Singh, S. (1998). Customer orientation and performance: a study of SMEs. *Management Decision*, 36, 385-394.

Archibugi, D., Cesaratto, S., & Sirilli, G. (1991). Sources of innovative activities and industrial organization in Italy. *Research Policy*, 20, 299-313.

Archibugi, D. & Coco, A. (2005). Measuring technological capabilities at the country level: A survey and a menu for choice. *Research Policy*, 34, 175-194.

Archibugi, D. & Pietrobelli, C. (2003). The globalisation of technology and its implications for developing countries: Windows of opportunity or further burden? *Technological Forecasting and Social Change*, 70, 861-883.

Armbruster, H., Kinkel, S., Lay, G. & Maloca, S. (2005). *Techno-organisational innovation in the European manufacturing industry. European Manufacturing Survey, Bulletin Nr.1.* Karlsruhe: Fraunhofer ISI.

Backes-Gellner, U., Maass, F., & Werner, A. (2005). On the explanation of horizontal, vertical and cross-sector R&D partnerships' evidence for the German industrial sector. *International Journal of Entrepreneurship and Innovation Management*, 5, 103-116.

Bacon, G., Beckman, S., Mowery, D. & Wilson, E. (1994). Managing product definition in high-technology industries. *California Management Review*, 36 (3), 32-56.

Baldwin, J. R. & Johnson, J. (1996). Business strategies in more-and less-innovative firms in Canada. *Research Policy*, 25, 785-804.

Barney, J. B. (1986). Strategic factor market: Expectations, luck and business strategy. *Management Science*, 32 (10) 1231-1241.

Barney, J. B. (1991). Firm resources and sustained competitive advantage. *Journal of Management*, 17, 99-120.

Barro, R. J. & Sala-i-Martin, X. (1995). *Economic growth.* New York: McGraw-Hill, Inc.

Barro, R. J. & Sala-i-Martin, X. (1997). Technological diffusion, convergence, and growth. *Journal of Economic Growth*, 2, 1-26.

Bassanini, A. & Scarpetta, S. (2001). The driving forces of economic browth: Panel data evidence for the OECD countries. *OECD Economic Studies*, 33. Retrieved November 18, 2008, from http:// http://hal.inria.fr/docs/00/16/83/83/PDF/EconomicStudies_01.pdf

Bayus, B. L. & Putsis, W. P. Jr. (1999). Product proliferation: An empirical analysis of product line determinants and market outcomes. *Marketing Science*, 18 (2), 137-153.

Belderbos, R., Carree, M. & Lokshin, B. (2004). Cooperative R&D and firm performance. *Research Policy*, 33, 1477-1492.

Belohlav, J. A. (1993). Quality, Strategy, and Competitiveness. *California Management Review*, 35, 55.

Bhidé, A. (2008). *The Venturesome Economy: How Innovation Sustains Prosperity in a More Connected World.* Princeton: Princeton University Press.

Bickart, B. A. (1993). Carryover and backfire effects in marketing research. *Journal of Marketing Research*, 30, 52-62.

Blundell, R., Griffith, R. & Van Reenen, J. (1999). Market share, market value and innovation in a panel of British manufacturing firms. *Review of Economic Studies*, 66, 529-554.

Bosworth, D. & Mahdian, H. (1999). Returns to intellectual property in the pharmaceuticals sector. *Economie Appliquee*, 52 (2), 69-93.

Bosworth, D. & Rogers, M. (2002). Market value, R&D and intellectual property: an empirical analysis of large Australian firms. *Economic Record*, 77 (239), 323-337.

Bougrain, F. & Haudeville, B. (2002). Innovation, collaboration and SMEs internal research capacities. *Research Policy*, 31 (5), 735-747.

Boyt, T. & Harvey, M. (1997). Classification of industrial services - a model with strategic implications. *Industrial Marketing Management*, 25, 291-300.

Bozdogan, K., Deyst, J., Hoult, D., & Lucas, M. (1998). Architectural innovation in product development through early supplier integration. *R&D Management*, 28, 163-173.

Brown, C. L. & Carpenter, G. S. (2000). Why is trivial important? A reasonsvased account for the effects of trivial attributes on choice. *Journal of Consumer Research*, 26 (4), 372-385.

Brown, S. L. & Eisenhardt, K. M. (1995). Product development: Past research, present findings, and future directions. *Academy of Management Review*, 20 (2), 343-378.

Buzzell, R. D. (2004). *The PIMS program of strategy research: A retrospective appraisal*. Journal of Business Research, 57, 478-483.

Calantone, R., Garcia, R. & Dröge, C. (2003). The effects of environmental turbulence on new product development strategy planning, *The Journal of Product Innovation Management*, 20 (2), 90-103.

Cassiman, B. & Veugelers, R. (2006). In search of complementarity in innovation strategy: Internal R&D and external knowledge acquisition. *Management Science*, 52, 68-82.

Caves, R. E. (1998). Industrial Organization and New Findings on the Turnover and Mobility of Firms. *Journal of Economic Literature*, 36, 1947-1982.

Cefis, E. & Marsili, O. (2006). Survivor: The role of innovation in firms' survival. *Research Policy*, 35, 626-641.

Celikel-Esser, F., Tarantola, S. & Mascherini, M. (2007). *Fourth European Community Innovation Survey: Strengths and weaknesses of European countries*. Institute for the Protection and Security of the Citizen. Retrieved November 21, 2008, from: http://statind.jrc.ec.europa.eu/Innovation/CIS_S&W-13Feb2007.pdf

Chaharbaghi, K. & Lynch, R. (1999). Sustainable competitive advantage: towards a dynamic resource-based strategy. *Management Decision*, 37, 45-50.

Chakrabarti, A., Singh, K. & Mahmodd, I. (2007). Diversification and performance: Ecvidence from East Asian firms, *Strategic Management Journal*, 28 (2), 101-120.

Chang, T. (1996). Cultivating global experience curve advantage on technology and marketing capabilities. *International Marketing Review*, 13, 22-42.

Chapman, R. & Hyland, P. (2004). Complexity and learning behaviors in product innovation. *Technovation*, 24, 553-561.

Chesbrough, H. W. (2003). The Era of Open Innovation. *MIT Sloan management review*, 44 (3), 35.
Chesbrough, H. W. & Garman, A. R. (2009). How open innovation can help you cope in lean times. *Harvard Business Review*, 87 (12), 68-76.
Chiesa, V., Giglioli, E. & Manzini, R. (1999). R&D Corporate Planning: Selecting the Core Technological Competencies. *Technology Analysis & Strategic Management*, 11 (2), 255-279.
Chiesa, V. & Manzini, R. (1997). Competence levels within firms: A static and dynamic analysis. In Heene, A. & Sanchez, R. (eds.). *Competence-based strategic management*. Chichester: Wiley, 195–214.
Chin, W. W. (1998). The partial least squares approach for structural equation modelling. In Marcoulides, G.A. (ed.), *Modern Methods for Business Research*. Hillsdale, NJ: Lawrence Erlbaum Associates, 295-336.
Chin, W. W., Marcolin, B. L.,& Newsted, P. N. (2003). A Partial Least Squares approach for measuring interaction effects: Results from a Monte Carlo simulation study and an electronic mail emotion/adoption study. *Information Systems Research*, 14 (2), 189-217.
Chin, W. W. & Newsted, P. R. (1999). Structural equation modeling analysis with small samples using partial least squares. In Hoyle, R. R. (ed.), *Statistical Strategies for Small Sample Research*. Thousand Oaks, CA: Sage, 307-341.
Clark, K. B. & Fujimoto, T. (1991). *Product development performance*. Boston: Harvard Business School Press.
Coates, T. T. & McDermott, C. M. (2002). An exploratory analysis of new competencies: a resource based view perspective. *Journal of Operations Management*, 20, 435-450.
Cohen, J. & Cohen, P. (1983). *Applied multiple regression/correlation analysis for the behaviorsl sciences*, 2nd edition. Hillsdale, NJ: Erlbaum.
Commission of the EC (2005). *Working together for growth and jobs: A new start for the Lisbon strategy*. Luxembourg: Office for Official Publications of the European Communities. Retrieved September 25, 2008, from http://eur-lex.europa.eu/ LexUriServ/LexUriServ.do?uri=COM:2005:0024:FIN:EN:PDF
Commission of the EC (2007). *Integrated guidelines for growth and jobs (2008-2010)*. Luxembourg: Office for Official Publications of the European Communities, Retrieved September 25, 2008, from http://ec.europa.eu/growthandjobs/pdf/european-dimension-200712-annual-progress-report/200712-annual-report-integrated-guidelines_en.pdf
Companies Act ZGD-1 (In Slovene: Zakon o gospodarskih družbah – ZGD-1). (2006). *Official gazette of Republic of Slovenia* (No. 42/2006).
Coombs, R. & Miles, I. (2000). Innovation, Measurement and Services: The New Problematique. In Metcalfe, J.S. & Miles, I. (eds.), *Innovation Systems in*

the Service Economy: Measurement and Case Study Analysis. Boston: Kluwer Academic Publishers, 85-103.

Cooper, R.G. & Kleinschmidt, E. J. (1995). Benchmarking the firm's critical success factors in new product development. *Journal of Product Innovation Management*, 12, 374-391.

Crosby, P. B. (1995). *Quality is still free: Making quality certain in uncertain times*. New York: McGraw-Hill.

Danneels, E. (2002). The dynamics of product innovation and firm competencies. *Strategic Management Journal*, 23 (12), 1095-1121.

Day, G. S. (1994). The capabilities of market-driven organizations. *Journal of Marketing*, 58 (4), 37-52.

DeFillippi, R. J. (1990). Casual ambiguity, barriers to imitation, and sustainable competitive advantage. *Academy of Management review*, 15, 88-102.

Deming, W. E. (2000). *Out of the Crisis*. Cambridge, MA: MIT Press.

Deshpandé, R. & Farley, J. U. (2004). Organizational culture, market orientation, innovativeness, and firm performance: an international research odyssey. *International Journal of Research in Marketing*, 21, 3-22.

Diamond, D. W. & Rajan, R. (2009). The credit crisis: Conjectures about causes and remedies. *NBER Working Paper Series*, Working Paper 14739. Retrieved February 9, 2011, from http://www.nber.org/papers/w14739

Dierickx, I. & Cool, K. (1989). Asset stock accumulation and the sustainability of competitive advantage. *Management Science*, 35 (12), 1504-1513.

Djellal, F. & Gallouj, F. (2001). Patterns of innovation organisation in service firms: postal survey results and theoretical models. *Science and Public Policy*, 28, 57-67.

Dolničar, S. (2003). Using cluster analysis for market segmentation – typical misconceptions, established methodological weaknesses and some recommendations for improvement. *Faculty of Commerce – Papers, University of Wollongong*. Retrieved August 20, 2008, from http://ro.uow.edu.au/commpapers/139

Dosch, F., Van den, A. J., Volberda, H. W. & de Boer, M. (1999). Coevolution of firm absorptive capacity and knowledge environment: Organizational forms and combinative capabilities. *Organization Science*, 10 (5), 551-568.

Drejer, I. (2004). Identifying innovation in surveys of services: A Schumpeterian perspective. *Research Policy*, 33, 551-562.

Drucker, P. F. (2007). *Innovation and Entrepreneurship*. Oxford: Butterworth-Heinemann.

Dumaine, B. (1989). How managers can succeed through speed. *Fortune*, 13, 54-59.

Duray, R. (2002). Mass customization origins: Mass or custom manufacturing? *International Journal of Operations and Production Management*, 22 (3), 314-328.

Easterly, W. & R. Levine (2001). What have we learned from a decade of empirical research on growth? It's not factor accumulation: Stylized facts and growth models. *The World Bank Economic Review*, 15 (2), 177-219.

Economist, The (2007). Out of the dusty labs: The rise and fall of corporate R&D. *The Economist*, 2007, March 3-9, 69-71.

Economist, The (2008). A gathering storm? *The Economist*, 2008, November 22-28, 67-68.

Eisenhardt, K. M. & Martin, J. A. (2000). Dynamic capabilities: What are they? *Strategic Management Journal*, 21, 1105-1121.

European Innovation Scoreboard (2008). PRO INNO Europe paper N° 6, Luxembourg: Office for Official Publications of the European Communities. Retrieved November 3, 2008, from http://www.proinno-europe.eu/admin/uploaded_documents/European_Innovation_Scoreboard_2007.pdf

Eurostat (2007). Fourth Community Innovation Survey. Brussels: Eurostat. Retrieved September 8, 2008, from http://epp.eurostat.ec.europa.eu/pls/portal/docs/page/pgp_prd_cat_prerel/pge_cat_prerel_year_2007/pge_cat_prerel_year_2007_month_02/9-22022007-en-bp.pdf

Ferligoj, A. (1989): *Razvrščanje v skupine (eng. Clustering). Metodološki zvezki, 4*. Ljubljana: Fakulteta za sociologijo, politične vede in novinarstvo.

Filippetti, A. & Archibugi, D. (2011). Innovation in times of crisis: National Systems of Innovation, structure, and demand. *Research Policy*, 40, 179-192.

Fisher, R.J. & Maltz, E. (1997). Enhancing communication between marketing and engineering: The moderating role of relative. *Journal of Marketing*, 61 (3), 54-70.

Flores, F. (1993). Innovation by listening carefully to customers. *Long Range Planning*, 26, 95-102.

Forbes, N. & Wield, D. (2000). Managing R&D in technology-followers. *Research Policy*, 29, 1095-1109.

Fornell, C. & Bookstein, F. L. (1982). Two structural equation models: LISREL and PLS applied to consumer exit-voice theory. *Journal of Marketing Research*, 19 (4), 440-452.

Fornell, C. & Cha, J. (1994). Partial least squares. In Bagozzi, R.P. (ed.), *Advanced Methods of Marketing Research* (52-78). Cambridge: Blackwell.

Fornell, C., Lorange, P. & Roos, J. (1990). The cooperative venture formation process: A latent variable structural modeling approach. *Management Science*, 36 (10), 1246-1255.

Fowler, S., King, A. W., Marsh, S. J., & Victor, B. (2000). Beyond products: new strategic imperatives for developing competencies in dynamic environments. *Journal of Engineering and Technology Management*, 17, 357-377.
Francis, D. & Bessant, J. (2005). Targeting innovation and implications for capability development. *Technovation*, 25, 171-183.
Fritsch, M. & Lukas, R. (2001). Who cooperates on R&D? *Research Policy*, 30, 297-312.
Fruman, C. C. F. (1992). Choices in R&D and business portfolio in the electronics industry. *Research Policy*, 21 (2), 97-124.
Frydman, R., Gray, C., Hessel, M. & Rapaczynski, A. (1999). When does privatization work? The impact of private ownership on corporate performance in the transition economies. *The Quarterly Journal of Economics*, 114 (4), 1153-1191.
Gallouj, F. & Weinstein, O. (1997). Innovation in services. *Research Policy*, 26, 537-556.
Garcia, R. & Calantone, R. (2002). A critical look at technological innovation typology and innovativeness terminology: A literature review. *Journal of Product Innovation Management*, 19 (2), 110-132.
Gassmann, O. & von Zedtwitz, M. (1999). New concepts and trends in international R&D organization. *Research Policy*, 28, 231-250.
Gottardi, G. (1996). Technology strategies, innovation without R&D and the creation of knowledge within industrial districts. *Industry & Innovation*, 3 (2), 119-134.
Gemser, G. & Leenders, M. (2001). How integrating industrial design in the product development process impacts on company performance. *Journal of Product Innovation Management*, 18, 28-38.
Geroski, P. A. (1995). What do we know about entry? *International Journal of Industrial Organization*, 13, 421-440.
Gertler, M. (2010). Banking crisis and real activity: Identifying the linkages. *International Journal of Central Banking*, 6, 4, 125-135.
Goffin, K. & New, C. (2001). Customer support and new product development. *International Journal of Operations & Production Management*, 21, 275-301.
Gollop, F. M. & Monahan, J. L. (1991). A Generalized Index of Diversification: Trends in US Manufacturing. *Review of Economics and Statistics*, 73, 318-330.
Grant, R. M. (1996). Prospering in Dynamically-Competitive Environments: Organizational Capability as Knowledge Integration. *Organizational Science*, 7, 375-387.
Grant, R. M. (2001). The resource-based theory of competitive advantage: Implications for strategy formulation. In Zack, M. H. (ed.), *Knowledge and Strategy*. Oxford: Butterworth-Heinemann, 3-24.

Greene, H. (2003). *Econometric Analysis*. New York: Prentice Hall.
Greenley, G. & Oktemgil, M. (1997). An investigation of modular effects on alignment skill, *Journal of Business Research*, 39 (2), 93–105.
Griffin, A. (1997). PDMA research on new product development practices: Updating trend and benchmarking best practices. *Journal of Product Innovation Management*, 14, 429-458.
Griffith, R., Redding, S., & Van Reenen, J. (2004). Mapping the Two Faces of R&D: Productivity Growth in a Panel of OECD Industries. *The Review of Economics and Statistics*, 86 (4), 883-895.
Griliches, Z. & Mairesse, J. (1983). Comparing productivity growth: An exploration of French and U.S. industrial firm data. *European Economic Review*, 21 (1-2), 89-119.
Grossman, G. M. & Helpman, E. (1994). Endogenous Innovation in the Theory of Growth. *The Journal of Economic Perspectives*, 8, 23-44.
Guellec, D. & van Pottelsberghe de la Potterie, B. (2001). *OECD Economic Studies*, (33). Retrieved October 12, 2008, from http:// http://www.oecd.org/dataoecd/26/32/1958639.pdf
Gupta, A. K. & Wilemon, D. L. (1990). Accelerating the development of technology-based new products. *California Management Review*, 32 (2), 24-44.
Gupta, A. K. & Wilemon, D. L. (1996). Changing patterns in industrial R&D management. *Journal of Product Innovation Management*, 13 (6), 497-511.
Hafeez, K., Malak N. & Zhang, Y. B. (2007). Outsourcing non-core assets and competences of a firm using analytic hierarchy process. *Computers & Operations Research*, 34 (12), 3592-3608.
Hafeez, K., Zhang, Y. B. & Malak, N. (2002). Core competence for sustainable competitive advantage: a structured methodology for identifying core competence. *Engineering Management, IEEE Transactions on*, 49, 28-35.
Hagedoorn, J. (2002). Inter-firm R&D partnerships: An overview of major trends and patterns since 1960. *Research Policy*, 31, 477-492.
Hagedoorn, J. & Cloodt, M. (2007). Measuring innovative performance: is there an advantage in using multiple indicators? *Research Policy*, 32, 1365-1379.
Hall, B.H. (1999). Innovation and market value. *NBER Working Paper Series*, w6984. Retrieved November 7, 2008, from http://www.nber.org/papers/W6984.pdf
Hall, L. A. & Bagchi-Sen, S. (2002). A study of R&D, innovation, and business performance in the Canadian biotechnology industry. *Technovation*, 22, 231-244.
Hall, R.W., Johnson, H.T. & Turney, P.B.B. (1991). *Measuring up: Charting pathways to manufacturing excellence*. Homewood: Business One Irwin.

Hamel, G. (1994). The Concept of Core Competence. In Hamel, G. & Heene, A. (eds.), *Competence-based Competition*. New York: Wiley, 11-33.

Handfield, R. G., Ragatz, G. L., Peterson, K. & Monczka, R. M. (1999). Involving suppliers in new product development? *California Management Review*, 42 (1), 59-82.

Hatcher, L. (1994). *A step-by-step approach to using the SAS system for factor analysis and structural equation modeling*. Cary: SAS Institute.

Hatziohronoglou, T. (1996) *Revision of the high-technology sector and product classification*. Paris: OECD.

Hay, D. A. & Morris, D. J. (1991). *Industrial economics and organization: Theory and evidence*. New York: Oxford University Press.

He, Z. L. & Wong, P. K. (2004). Exploration vs. exploitation: An empirical test of the ambidexterity hypothesis. *Organization Science*, 15, 481-494.

Hertenstein, J. H., Platt, M. B., & Veryzer, R. W. (2005). The Impact of Industrial Design Effectiveness on Corporate Financial Performance. *Journal of Product Innovation Management*, 22, 3-21.

Hill, C. W. L. (1988). Differentiation Versus Low Cost or Differentiation and Low Cost: A Contingency Framework. *Academy of Management Review*, 13, 4.

Hirsch-Kreinsen, H., Jacobson, D., & Robertson, P. L. (2006). 'Low-tech' Industries: Innovativeness and Development Perspectives - A Summary of a European Research Project. *Prometheus*, 24, 3-21.

Hitt, M. A. & Ireland, R. D. (1985). Corporate distinctive competence, strategy, industry and performance. *Strategic management journal*, 6, 273-293.

Hitt, M. A., Ireland, R. D., Rowe, G. & Sheppard, J. (2005). Chapter four: The Internal environment: Resources, capabilities, and core competencies. In Hitt, M. A., Duane Ireland, R. & Hoskisson, R. E. (eds.), *Strategic management: Concepts: Competitiveness and globalization*, 2nd Canadian Edition. Toronto: Nelson College Indegenous.

Hollensen, S. (2007). *Global marketing: A decision-oriented approach*. Essex: Pearson Education Limited.

Hughes, A. & Wood, E. (1999). Rethinking innovation comparisons between manufacturing and services: The experience of the CBR SME survey in the UK. In Metcalfe, J.S. & Miles, I. (eds.), *Innovation Systems in the Service Economy: Measurement and Case Study Analysis* (105-124). Boston: Kluwer Academic Publishers.

Hult, G.T.M., Hurley, R.F. & Knight, G.A. (2004). Innovativeness: Its antecedents and impact on business performance. *Industrial Marketing Management*, 33, 429-438.

Ivarsson, I. & Jonsson, T. (2003). Local technological competence and asset-seeking FDI: An empirical study of manufacturing and wholesale affiliates in Sweden. *International Business Review*, 12 (3), 369-386.

Jaruzelski, B. & Dehoff, K. (2008). Beyond Borders: The Global Innovation 1000. *Strategy + business*, issue 53, Winter 2008: Booz & Company.
Jeong, I., Pae, J.H. & Zhou, D. (2006). Antecedents and consequences of the strategic orientations in new product development: The case of Chinese manufacturers. *Industrial Marketing Management*, 35, 348-358.
Juran, J. M. (1989). *Leadership for Quality: An Executive Handbook*. New York: Free Press.
Kahurana, A. & Rosenthal, S. R. (1997). Integrating the fuzzy end of new product development. *MIT Sloan Management Review*, 38 (2), 103-120.
Kano, N., Seraku, N., Takahashi, F. & Tsuji, S. (1984). Attractive quality and must-be quality. *The Journal of the Japanese Society for Quality Control*, 14, 39-48.
Kanter, R. M. (2006). Innovation: The classic traps. *Harvard Business Review*, 84 (11), 72-83.
Kim, J.Y., Wong, V. & Eng, T.Y. (2005). Product variety strategy for improving new product development proficiencies. *Technovation*, 25, 1001-1015.
Kirner, E., Kinkel, S., & Jaeger, A. (2009). Innovation paths and the innovation performance of low-technology firms - An empirical analysis of German industry. *Research Policy*, 38 (3), 447-458.
Klenow, P. J. & Rodriguez-Clare, A. (1997). The neoclassical revival in growth economics: Has it gone too far? *NBER Macroeconomics Annual*, 73-103.
Klevorick, A. K., Levin, R. C., Nelson, R. R. & Winter, S. G. (1995). On the sources and significance of inter-industry differences in technological opportunities. *Research Policy*, 24 (2), 185-205.
Koen, P. A. & Kohli, P. (1998). Idea Generation: Who Has the Most Profitable Ideas. *Engineering Management Journal*, 10, 35-40.
Kogut, B. & Zander, U. (1992). Knowledge of the Firm, Combinative Capabilities, and the Replication of Technology. *Organization Science*, 3, 383-397.
Kok, W. (2004). Facing the challenge: The Lisbon strategy for growth and employment. *Report from the High Level Group*, Luxembourg: Office for Official Publications of the European Communities. Retrieved November 8, 2008, from
http://www.umic.pt/images/stories/publicacoes200801/kok_report_en.pdf
Koufteros, X. & Marcoulides, G.A. (2006). Product development practices and performance: A structural equation modeling-based multi-group analysis. *International Journal of Production Economics*, 103, 286-307.
Kroll, M., Wright, P., & Heines, R. A. (1999). The contribution of product quality to competitive advantage: Impacts on systematic variance and unexplained variance in returns. *Strategic management journal*, 20, 375-384.
Kumiko, M. (1994). Interlinkages between systems, key components and component generic technologies in building competencies. *Technology Analysis and Strategic Management*, 6 (1), 107-120.

Langerak, F., Peelen, E. & Commandeur, H. (1997). Organizing for effective new product development: An exploratory study of Dutch and Belgian industrial firms. *Industrial Marketing Management*, 26 (3), 281-289.
Lanjouw, J. O. & Schankerman, M. A. (1999). The Quality of Ideas: Measuring Innovation with Multiple Indicators. *NBER Working Paper Series*, w7345. Retrieved November 7, 2008, from http://ssrn.com/abstract=194848
Leifer, R., Colarelli O'Connor, G. & Rice, M. (2001). Implementing radical innovation in mature firms: The role of hubs. *Academy of Management Executive*, 15 (3), 102-113.
Leonard-Barton, D. (1992). Core Capabilities and Core Rigidities: A Paradox in Managing New Product Development. *Strategic Management Journal*, 13, 111-125.
Leong, F. T. L.& Austin, J. T. (2006). *The Psychology Research Handbook: A Guide for Graduate Students and Research Assistants*. London: SAGE.
Li, T. & Calantone, R. J. (1998). The Impact of Market Knowledge Competence on New Product Advantage: Conceptualization and Empirical Examination. *Journal of Marketing*, 62, 13-29.
Li, T. & Cavusgil, S. T. (2000). Decomposing the Effects of Market Knowledge Competence in New Product Export. *European Journal of Marketing*, 34, 57-79.
Lichtenberg, F. R. & Siegel, D. (1991). The impact of R&D investment on productivity – new evidence using linked R&D-LRD data. *Economic Inquiry*, 29, 203-229.
Lokshin, B., Van Gils, A. & Bauer, E. (2009). Crafting firm competencias to improve innovative performance. *European Management Journal*, 27 (3), 187-196.
Lucas, R. E. Jr. (1988). On the mechanics of economic development. *Journal of Monetary Economics*, 22, 3-42.
Luchs, B. (1990). Quality as a strategic weapon. *European Business Journal*, 2 (4), 34-47.
Lukas, B. A. & Ferrell, O. C. (2000). The effect of market orientation on product innovation. *Journal of the Academy of Marketing Science*, 28, 239-247.
Lynn, G. S., Morone, J. G. & Paulson, A. S. (1996), Marketing and discontinuous innovation. *California Management Review*, 38 (3), 8-37.
Lynskey, M. J. (1999). The Transfer of Resources and Competencies for Developing Technological Capabilities - The Case of Fujitsu-ICL. *Technology Analysis and Strategic Management*, 11, 317-336.
Mairesse, J. & Sassenou, M. (1991). R&D productivity: A survey of econometric studies at the firm level, *NBER Working Paper Series*, w3666. Retrieved November 8, 2008, from http://www.nber.org/papers/w3666.pdf

Malerba, F. (1993). The National System of Innovation: Italy. In Nelson, R. R. (ed.), *National innovation systems: a comparative analysis*. New York: Oxford University Press, 230-260.

Malhotra, N. K. & Birks. D. F. (2003). *Marketing research: An applied approach*. London: Prantice-Hall, Inc.

Mankiew, N. G. (2003). *Macroeconomics*, 5th edition. New York: Worth Publishers.

Mansfield, E. (1984). Chapter 6 - R&D and Innovation: Some empirical findings. In Griliches, Z. & Pakes A. (eds.), *Patents, R&D and Productivity*. Chicago: University of Chicago Press.

Mansfield, E. & Wagner, S. (1975). Organizational and strategic factors associated with probabilities of success in industrial R&D, *The Journal of Business*, 48, 179-198.

Markides, C. C. & Williamson, P. J. (1994). Related diversification, core competencies and corporate performance. *Strategic Management Journal*, 15, 149-165.

Marsili, O. & Salter, A. (2006). The dark matter of innovation: Design and innovative performance in Dutch manufacturing. *Technology analysis and strategic management*, 18 (5), 515-534.

Mazzoleni, R. & Nelson R. R. (1998). The benfits and costs of strong patent protection: A contribution to the current debate. *Research Policy*, 27 (3), 273-284.

McEvily, S. K., Eisenhardt, K. M. & Prescott, J. E. (2004). The Global Acquisition, Leverage, and Protection of Technological Competencies. *Strategic Management Journal*, 25, 713-722.

McKinsey&Company (2010). Innovation and commercialization. *McKinsey Quarterly*, August 2010. Retrieved February 9, 2011 from https://www.mckinseyquarterly.com/Innovation_and_commercialization_2010_McKinsey_Global_Survey_results_2662

Menon, A., Chowdhury, J., & Lukas, B. A. (2002). Antecedents and outcomes of new product development speed: An interdisciplinary conceptual framework. *Industrial Marketing Management*, 31, 317-328.

Metacalfe, S. (1995). The Economic Foundations of Technology Policy: Equilibrium and Evolutionary Perspectives. In Stoneman, P. (ed.), *Handbook of the Economics of Innovation and Technological Change* (409-512). Oxford: Blackwell Publishers.

Mikkola, J. H. (2001). Portfolio management of R&D projects: implications for innovation management. *Technovation*, 21, 423-435.

Ministry of Economic Affairs. (1996). *The Dutch SME Sector: An International Comparison*. Hague: Ministry of Economic Affairs.

Montgomery, C. A. (1994). Corporate diversification. *Journal of Economic Perspectives*, 8 (3), 163-178.

Morone, J. (1993). *Winning in high-tech markets: The role of general management.* Boston, Massachusetts: Harvard Business School Press.
Nagaoka, S. (2006). R&D and market value of Japanese firms in the 1990s. *Journal of the Japanese and International Economies,* 20, 155-176.
Napolitano, G. (1991). Industrial Research and Sources of Innovation: A Cross-Industry Analysis of Italian Manufacturing Firms. *Research Policy,* 20, 171-178.
Nelson, R. R. (1993). National innovation systems: A retrospective. New York: Oxford University Press. In Nelson, R. R. (ed.), *National innovation systems: A comparative analysis.* New York: Oxford University Press, 505-524.
Nelson, R. R. & Rosenberg, N. (1993). Technical Innovation and National Systems. In Nelson, R. R. (ed.), *National innovation systems: A comparative analysis.* New York: Oxford University Press, 3-22.
Nelson, R. R. & Winter, S. G. (1982). *An evolutionary theory of economic change.* Boston, MA: The Belknap Press of Harvard University Press.
OECD (2010). OECD Science, technology and industry outlook 2010 – highlights. *OECD Outlook 2010.* Retrieved February 9, 2011 from http://www.oecd.org/dataoecd/38/13/46674411.pdf.
OECD (2006). *Community Innovation Statistics* Retrieved August 6, 2008, from http://www.oecd.org/dataoecd/37/39/37489901.pdf
OECD. (1997). *Revision of High Technology Sector and Product Classification. STI Working Papers 1997/2.* Paris: OECD.
OECD/Eurostat. (1997). *Proposed guidelines for collecting and interpreting technological innovation data - Oslo manual.* Paris: OECD.
Patel, P. & Pavitt, K. (1995). Patterns of technological activity: their measurement and interpretation. In Stoneman, P. (ed.), *Handbook of the Economics of Innovation and Technological Change* (14-51). Oxford: Blackwell Publishers.
Paul, J. V. & Peter, J. W. (1994). Core competencies, competitive advantage and market analysis: forging the links. In Hamel, G. & Heene, A. (eds.), Competence-based Competition (77-110). New York: Wiley.
Pavitt, K. (1990). What we know about the strategic management of technology. *California Management Reviw,* 32, 17-26.
Penrose, E. T. & Pitelis, C. (2009). *The theory of the growth of the firm.* London: Oxford University Press.
Peteraf, M. A. (1993). The Cornerstones of Competitive Advantage: A Resource-Based View. *Strategic Management Journal,* 14, 179-191.
Porter, M. E. (1998a). *Competitive advantage.* New York: Free Press.
Porter, M. E. (1998b). *Competitive strategy: Techniques for analyzing industries and competitors.* New York: Free Press.
Porter, M. E. (1999) Creating Advantage. Executive Excellence, 16 (11), 13-14.

Prahalad, C. K. & Hamel, G. (1990). The core competence of the corporation. *Harvard Business Review*, 68, 79-91.
Prajogo, D. I., McDermott, P., & Goh, M. (2008). Impact of value chain activities on quality and innovation. *International Journal of Operations & Production Management*, 28 (7), 615-635.
Prasad, B. (1996). *Concurrent Engineering Fundamentals, Volume II: Integrated Product Development*. Upper Saddle River, NJ: Prentice-Hall PTR.
Prašnikar, J. (2006). From Lisbon to "Lisbon". In Prašnikar, J. (ed.), *Competitiveness, social responsibility and economic growth* (2-18). Hauppague: Nova Science.
Prašnikar, J., Lisjak, M., Rejc Buhovac, A. & Štembergar, M. (2008). A methodology for analyzing inter-relationships between technological and marketing capabilities. *Long Range Planning*, 41 (5), 530-554.
Priem, R. L. & Butler, J. E. (2001). Is the resource-based" view" a useful perspective for strategic management research. *The Academy of Management Review*, 26 (1), 22-40.
Ramanujam, V. & Varadarajan, P. (1989). Research on corporative diversification: A synthesis. *Management Journal*, 10 (6), 523-551.
Ravald, A. & Grönroos. C. (1996). The value concept and relationship marketing. *European Journal of Marketing*, 30 (2), 19-30.
Rebelo, S. (1991). Long-run policy and long-run growth. *Journal of Political Economy*, 99 (3), 500-521.
Reeves, M. & Deimler, M. S. (2009). Strategies for winning in the current and post-recession environment. *Strategy & Leadership*, 37, 6, 10-17.
Reich, R. (1991). *The Work of Nations*. New York: Vintage.
Ringle, C.M., Wende, S. & Will, A. (2005). SmartPLS 2.0 (beta). Hamburg: SmartPLS. Retrieved August 30, 2008, from http://www.smartpls.de
Reinhart, C. M. & Rogoff, K. S. (2009). The aftermath of financial crises. *NBER Working Paper Series*, Working Paper 14656. Retrieved February 9, 2011, from http://www.nber.org/papers/w14656
Ritchie, L. & Dale, B. G. (1999). Self-assessment the business excellence model: A study of practice and process. *International Journal of Production Economics*, 66, 241-254.
Ritter, T. & Gemunden, H. G. (2004). The impact of a company's business strategy on its technological competence, network competence and innovation success. *Journal of Business Research*, 57, 548-556.
Robertson, P. L. & Patel, P. R. (2007). New wine in old bottles: Technological diffusion in developed economies. *Research Policy*, 36, 708-721.
Robertson, P. L., Smith, K., & von Tunzelmann, N. (2009). Innovation in low- and medium-technology industries. *Research Policy*, 38 (3), 441-446..
Romer, P. M. (1986). Increasing returns and long-run growth. *Journal of Political Economy*, 94, 1002-1037.

Rothaermel, F. T. (2001). Research note: Incumbent's Advantage Through Exploiting Complementary Assets via Interfirm Cooperation. *Strategic Management Journal*, 22, 687-699.
Rumelt, R. P. (1984). Towards a strategic theory of the firm. In Lamb, R. B. (ed.) *Competitive Strategic Management*. Englewood Cliffs, NJ: Prentice Hall, 556-570.
Rumelt, R. P. (1991). How much does industry matter? *Strategic Management Journal*, 12, 167-185.
Rumelt, R. P. (1994). Foreword. In Hamel, G. & Heene, A. (eds.), Competence-based Competition. New York: Wiley.Sanchez, R. (1995). Strategic flexibility in product competition. *Strategic Management Journal*, 16, 135-159.
Sanchez, R. (1995). Strategic flexibility in product competition. *Strategic Management Journal*, 16, 135-159.
Sanchez, R. (2002). Building blocks for strategy theory: Resources, dynamic capabilities and competences. In Volberda, H. W. & Elfring, T. (eds.), *Rethinking Strategy*. London: Sage Publications, 143-157.
Sanchez, R. (2004). Understanding competence-based management Identifying and managing five modes of competence. *Journal of Business research*, 57, 518-532.
Sanchez, R. & Heene, A. (1997). Reinventing Strategic Management: New Theory and Practice for Competence-based Competition. *European Management Journal*, 15, 303-317.
Sanchez, R., Heene, A. & Thomas, H. (1996). Towards the theory and practice of competence-based competition. In Sanchez, R., Heene, A. & Thomas, H. (eds.), *Dynamics of competence-based competition: Theory and practice in the new strategic management*. London: Elsevier, 1-35.
Sandven, T., Smith, K., & Kaloudis, A. (2005). Structural change, growth and innovation: the roles of medium and low-tech industries, 1980-2000. In Hirsch-Kreinsen, H., Jacobson, D. & Laestadius, S. (eds.), *Low-Tech Innovation in the knowledge economy*. Frankfurt-am-Main: Peter Lang, 31-59.
Sapir, A. (2003). *An agenda for a growing Europe*: Making the EU economic system deliver. *Report of an Independent High-Level Study Group established on the initiative of the President of the European Commission*. Retrieved November 7, 2008, from http://www.euractiv.com/ndbtext/innovation/sapirreport.pdf
Schewe, G. (1994). Successful innovation management: An integrative perspective. *Journal of Engineering and Technology Management*, 11, 25-53.
Schewe, G. (1996). Imitation as a strategic option for external acquisition of technology. *Journal of Engineering and Technology Management*, 13, 55-82.
Schmalensee, R. (1985). Do markets differ much? *American Economics Review*, 75 (3), 341-351.

Schmalensee, R. (1988). Inter-industry studies of structure and performance. In Schmalensee, R. & Willig, R. D. (eds.), *Handbook of Industrial Organization*, 2nd edition, Amsterdam: North Holland.
Schumpeter, J. A. (1983). *The Theory of Economic Development: An Inquiry Into Profits, Capital, Credit, Interest, and the Business Cycle.* Piscataway: Transaction Publishers.
Sirilli, G. & Evangelista, R. (1998). Technological innovation in services and manufacturing: results from Italian surveys. *Research Policy*, 27 (9), 881-899.
Slater, S. F. & Narver, J. C. (1994). Market Orientation, Customer Value, and Superior Performance. *Business Horizons*, 37, 22-28.
Solow, R. M. (1956). A contribution to the theory of economic growth. *The Quarterly Journal of Economics*, 70 (1), 65-94.
Solow, R. M. (1957). Technical change and the aggregate production function. *Review of Economics and Statistics*, 39 (3), 312-320.
Song M., Droge C., Hanvanich S. & Catalone R. (2005). Marketing and Technology Resource Complementarity: An Analysis of Their Interaction Effect in Two Environmental Contexts. *Strategic Management Journal*, 26, 259-276.
Song, X. M., Thieme, R. J. & Xie, J. (1998). The impact of cross-functional integration across product development stages: An exploratory study. *Journal of Product Innovation Management*, 15 (4), 289-303.
Stanovnik, P. & Kos, M. (2005). *Technology foresight in Slovenia.* Working Paper No. 27, Ljubljana: Institute for Economic Research. Retrieved November 8, 2008, from http://www.ier.si/files/Working%20paper-27.pdf
Storey, C. & Easingwood, C. J. (1998). The augmented service offering: a conceptualization and study of its impact on new service success. *Journal of Product Innovation Management*, 15, 335-351.
Strategic Planning Institute, The. (2008). *The PIMS Database.* The Strategic Planning Institute. Retrieved November 16, 2008, from http://www.pimsonline.com/pims-db.htm
Sundbo, J. & Gallouj, F. (2000). Innovation as a loosely coupled system in services. *International Journal of Services Technology and Management*, 1, 15-36.
Swink, M. & Song, M. (2007). Effects of marketing-manufacturing integration on new product development time and competitive advantage. *Journal of Operations Management*, 25, 203-217.
Tabachnick, B.G. & Fidell, L.S. (2007). *Using Multivariate Statistics.* (5th ed.) New York: Pearson Education Inc.
Teece, D. J. (1980). Economies of scope and the scope of the enterprise. *Journal of Economic Behavior and Organization*, 1, 223-247.

Teece, D. J. (1988). Capturing value from technological innovation: Integration, strategic partnering, and licensing decisions. *Interfaces*, 18 (3), 46-61.
Teece, D. J., Pisano, G., & Shuen, A. (1997). Dynamic capabilities and strategic management. *Strategic Management Journal*, 18, 509-533.
Tenenhaus, M., Esposito Vinzi, V., Chatelin, Y.M. & Lauro, C. (2005). PLS path modeling. *Computational Statistics and Data Analysis*, 48, 159–205.
Tether, B. S. (2002). Who co-operates for innovation, and why-An empirical analysis. *Research Policy*, 31, 947-967.
Tether, B. S., Hipp, C. & Miles, I. (2001). Standardisation and particularisation in services: evidence from Germany. *Research Policy*, 30, 1115-1138.
Thomke, S. & von Hippel, E. (2002). Customers as innovators, a new way to create value. *Harvard Business Review*, 80 (4), 73-81.
Thompson, A. A., Strickland, A. J., & Gamble, J. E. (2009). *Crafting and executing strategy: The quest for competitive advantage – concepts and cases (17th International Edition)*. New York: McGraw-Hill/Irwin.
Tidd, J. (2006). The competence cycle: Translating knowledge into new processes, products and services. In Tidd, J. (ed.), *From knowledge management to strategic competence*. Singapore: Imperial College Press.
Tidd, J., Bessant, J. R., & Pavitt, K. (1997). *Managing innovation: Iintegrating technological, market and organizational change*. New York: Wiley.
Tidd, J. & Bodley, K. (2002). The influence of project novelty on the new product development process. *R&D Management*, 32 (2), 127-138.
Tidd, J., Driver, C. & Saunders, P. (1996). Linking technological, market and financial indicators of innovation. *Economics of Innovation and New Technology*, 4 (3), 155-172.
Tödtling, F., Lehner, P., & Kaufmann, A. (2009). Do different types of innovation rely on specific kinds of knowledge interactions? *Technovation*, 29, 59-71.
Tong, J. & Xu, C. (2006). How European financial institutions affect their R&D and economic development. In Prasnikar, J. (ed.), *Competitiveness, Social Responsibility and Economic Growth*. Hauppauge NY: U.S.A., Nova Science Publishers, Inc., 31-47.
Torkkeli, M. & Tuominen, M. (2002). The contribution of technology selection to core competencies. *International Journal of Production Economics*, 77, 271-284.
Treacy, M. & Wiersima, F. (1993). Customer Intimacy and Other Value Disciplines. *Harvard Business Review*, 71 (1), 84-93.
Tuominen, M., Moller, K. & Rajala, A. (1997). Marketing capability: a nexus of learning-based resources and a Prerequisite for Marketing Orientation. In Arnott, D., Bridgewater, S., Dibb, S., Doyle, P., Freeman, J., Melewar, T., Shaw, V., Simkin, L., Stern, P., Wensley, R. & Wong, V. (eds.), *Marketing: Progress, Prospects, Perspectives, Warwick, 26th EMAC Conference Proceedings, Volume 3* (1220-1240). UK: Warwick Business School.

Tyler, B. B. (2001). The complementarity of cooperative and technological competencies: A resource-based perspective. *Journal of Engineering and Technology Management*, 18 (1) 1-27.

Varadarajan, R. (2009). Fortune at the bottom of the innovation pyramid: the strategic logic of incremental innovations. *Business Horizons*, 52 (1), 21-29.

Verdin, P. J. & Williamson, P. J. (1991). From barriers to entry to barriers to survival. A paper presented at the 11th Annual International Conference of the Strategic Management Society, Toronto, October 1991.

Veryzer, R. W. (1998). Key Factors Affecting Customer Evaluation of Discontinuous New Products. *Journal of Product Innovation Management*, 15, 136-150.

Veryzer, R. W. (2005). The Roles of Marketing and Industrial Design in Discontinuous New Product Development. *Journal of Product Innovation Management*, 22, 22-41.

von Tunzelmann, N. & Acha, V. (2005). Innovation in "Low-tech" Industries. In Fagerberg, J., Mowery, D.C. & Nelson R.R. (eds.), *The Oxford Handbook of Innovation*. New York: Oxford University Press, 407-432.

von Zedtwitz, M. (2004). Managing foreign R & D laboratories in China. *R&D Management*, 34, 439-452.

Vorhies, D. W. (1998). An investigation of the factors leading to the development of marketing capabilities and organizational effectiveness. *Journal of Strategic Marketing*, 6, 3-23.

Vorhies, D. W. & Harker, M. (2000). The Capabilities and Performance Advantages of Market-Driven Firms: An Empirical Investigation. *Australian Journal of Management*, 25 (2), 145-172.

Vorhies, D. W., Harker, M. & Rao, C. P. (1999). The capabilities and performance advantages of market-driven firms. *European Journal of Marketing*, 33, 1171-1202.

Wakelin, K. (2001). Productivity growth and R&D expenditure in UK manufacturing firms. *Research Policy*, 30, 1079-1090.

Walsh, S. & Linton, J. D. (2002). The measurement of technical competencies. *The Journal of High Technology Management Research*, 13, 63-86.

Wang, Y., Lo, H. P. & Yang, Y. (2004). The Constituents of Core Competencies and Firm Performance: Evidence from High-technology Firms in China. *Journal of Engineering and Technology Management*, 21, 249-280.

Ward, J. H. (1963). Hierarchical grouping to optimize an objectivefunction. *Journal of the American Statistical Association*, 58, 236-244.

Wernerfelt, B. (1984). A Resource-Based View of the Firm. *Strategic Management Journal*, 5, 171-180.

Wheelwright, S. C. & Clark, K. B. (1992). *Revolutionizing Product Development: Quantum Leaps in Speed, Efficiency, and Quality*. New York: Free Press.

White, G.I., Sondhi, A.C. & Fried, D. (2003). *The analysis and use of financial statements*. Hoboken: John Wiley & Sons, Inc.

Wold, H. (1981). *The fix-point approach to interdependent systems: Review and current outlook*. Amsterdam: North Holland. In Wold, H. (ed.), *The fix-point approach to interdependent systems*. Amsterdam: North-Holland, 1-35.

Wold, H. (1982). Soft modelling: The basic design and some extensions. In Jöreskog, K. G., Wold, H. (eds), *Systems under Indirect Observation, Part 2*. Amsterdam: North Hollad, 1-54.

Young, A. (1995). The tyranny of numbers: Confronting the statistical realities of the East Asian growth experience. *Quarterly Journal of Economics*, 110 (3), 641-680.

Zahra, S.A. & Ellor, D. (1993). Accelerating new product development and successful market introduction. *SAM Academy of Management Journal*, 58 (1), 9-15.

Zirger, B. J. & Maidique, M. A. (1990). A Model of New Product Development: An Empirical Test. *Management Science*, 36, 867-883.

APPENDICES

Appendix A

Comparison of the contemporary strategic management approaches

	Resource-based theory (1980's)	Dynamic capabilities theory (1990's)	Competence-based theory (1990's)
Concept of a firm	A bundle of resources and capabilities comprising:	A system formed by processes, routines and resources comprising:	An open system of asset stocks and flows comprising:
		• Tangible assets • Intangible assets • Capabilities	
	Activities	Organisational/ managerial process	Managerial process
Competitive strategy	Controlling and exploiting strategic resources manifested in assets or capabilities	Deploying and exploiting capabilities embedded in processes, and continually reshaping of the portfolio of assets	Deploying, protecting and developing competencies resulted from the integration of assets and capabilities
Attributes of resources/ competencies		• Valuable • Rare • Inimitable • Non-substitutable	
		Dynamic	Robust (for new market)
Development method	Development of intangible assets	Development and integration of intangible assets and capabilities	
Development environment	Internal only	Internal and external	

Sources: Wernerfelt (1984), Prahalad & Hamel (1990), Hamel (1994), Sanchez & Heene (1997), Teece et al. (1997)

Appendix B

Studies aimed at developing the theory of competencies

Authors: Lokshin, Gils, Bauer (2009)

Concepts used:
- Customer, technological and organizational competencies
- Innovative performance

Methodology:
- Structured questionnaire
- 27 German firms from the fast moving consumer goods industry
- Factor analysis, multivariate regression analysis
- Organizational competencies measured with two indicators: team structure and slack time

Findings:
- Confirmed direct effect of organizational competencies on innovative performance
- Synergetic effect of combining technological, customer and organizational competencies on product innovation, especially key for radical innovation
- Higher levels of competencies are characteristic of firms with higher innovation output
- Radical innovations require higher levels of firm competencies than incremental innovations

Authors: Ritter, Gemünden (2004)

Concepts used:
- Business strategy
- Technological and network competence
- Innovation success

Methodology:
- 308 German companies in mechanical and electrical engineering
- SEM using LISREL (7 point scale)

Findings:
- Technological and network competence (strategic flexibility) both affect innovation success
- Business strategy (limited to technology) is not directly related to innovation success but supports development of both competencies
- Industry specific (environmental characteristics) not included

Authors: Wang, Lo, Yang (2004)

Concepts used:
- Marketing, technological and integrative competencies
- Environmental turbulence: market and technological turbulence
- Integrated firm performance

Methodology:
- Stratified sample of 248 high-tech firms in China
- SEM using PLS and evaluation of main effects

Findings:
- Marketing, technological and integrative competencies have significant influences on firm performance
- Relationships significantly moderated by environmental turbulence; market turbulence has no effect on the relationship between integrative competencies and firm performance

Authors: Coates, McDermott (2002)

Concept used:
- Technology, market and integration competencies

Methodology:
- Within-case analysis (longitudinal study based on interviews) of an emerging technology project of a large US high-tech manufacturing company
- Exploratory qualitative analysis

Findings:
- Observed 3 groups of newly generated competencies that supported the development of emerging technology: technology, market and integration competencies
- Competencies are complex skill sets acquired through learning that have to be managed
- Within groups described the role of specific abilities and assets for the success of the project and firm as a whole
- New competencies help firm develop attractive product market positions and gain advantages as a first mover

Authors: Fowler, King, Marsh, Victor (2000)

Concepts used:
- Market-driven, technological and integration competencies
- Dynamic environments

Methodology:
- Propositions built on the synthesis of existing literature from theory and practice

Findings:
- Exploiting new opportunities through competencies instead of products
- By focusing on competencies companies place less emphasis on product-centred strategies
- Strategies based on competencies are superior to product-centred strategies in dynamic environments
- Competencies are associated with competitive advantage in dynamic environments

Author: Chang (1996)

Concepts used:
- Technology and marketing capability
  Note: Author used the term capability, however, the description of the concept corresponds to the definition of competence chosen to be used in the thesis
- Profitability and performance

Methodology:
- PIMS database: 2744 firms from USA, Canada, UK, EU
- 52% market pioneers, 48% market followers and late entrants
- 28% consumer product business, 72% industrial product manufacturers
- OLS regression; use of 5- and 3-point nominal scales, interval scale and ratio scale

Findings:
- Market pioneers possess significantly higher technology and marketing capabilities than market followers
- Technology and market capabilities contribute significantly to the firm's ROI, ROS, cash flow on investment (CFL/Invest) and market share
- Interaction between technology and marketing capabilities exists with respect to market share (no explanation provided why interaction effect does not have a significant influence on ROI, ROS and CFL/Invest)
- Selection of indicators limited by the number of indicators included in the PIMS database
- Framework for developing global experience curve advantage – technology and marketing capabilities are two basic dimensions of a firm's global learning

Authors: Hitt, Ireland (1985)

Concepts used:
- Corporate distinctive competencies (general administration, production/operation, engineering and R&D, marketing, finance, personnel, public and governmental relations)
- Firm performance
- Grand strategies (stability, internal growth, external acquisitive growth and retrenchment)
- 4 industries (consumer non-durable/durable goods, capital goods, producer goods)

Methodology:
- 185 Fortune 1000 industrial firms
- Moderated regression analysis

Findings:
- Corporate distinctive competencies affect firm performance
- Strategy and industry act as moderators

Appendix C

Questionnaire:
Competencies and innovative performance of firms

Company name	Company ID number

Since what year has the company been present in the industry (regardless of changes of the organizational form)

Does your company belong to a group of companies? ☐ Yes ☐ No

A. BASIC PRODUCT RELATED QUESTIONS

A.1. Please name main product lines of your company and their market shares:

Name of the product line	Share of sales during period 2005-2007
	%
	%
	%
	%
Independent services[3]	%

---

[3] Services not directly related to own products, such as for example representation and sale of foreign products, repair shop, etc.

B. QUESTIONS RELATING TO INDUSTRY AND FIRM COMPETENCIES

(i) If you believe that among product lines listed under question A.1., there are few differences with respect to market in firm capabilities involved in their development and marketing, please provide answers for the **company as a whole**. In a different case provide answers for the **main product line** or individually for the **more representative** ones.

B.1. Answers to be provided for:
☐ **Company as a whole** (all product lines together)
☐ **Individual product lines** (chosen product lines in table A. 1. mark by letters A, B and C)

B.2. To what extent do you agree with the following statements about the industry and market?
To answer use the following scale:
1 – Strongly disagree, 2 – Disagree, 3 – Neither agree nor disagree, 4 – Agree, 5 – Strongly agree

	Whole company or A					B					C				
The level of market uncertainty is extremely high.	1	2	3	4	5	1	2	3	4	5	1	2	3	4	5
It is almost impossible to predict accurately the rapidly changing tastes and demands of consumers.	1	2	3	4	5	1	2	3	4	5	1	2	3	4	5
Activities of major competitors are unpredictable.	1	2	3	4	5	1	2	3	4	5	1	2	3	4	5
The competition in the industry is very intense.	1	2	3	4	5	1	2	3	4	5	1	2	3	4	5
Technology[4] in our industry is changing rapidly.	1	2	3	4	5	1	2	3	4	5	1	2	3	4	5
New technologies have a high impact on business operations and competition and bring about big opportunities.	1	2	3	4	5	1	2	3	4	5	1	2	3	4	5
It is very difficult to predict technological changes in the next 2 to 3 years.	1	2	3	4	5	1	2	3	4	5	1	2	3	4	5
Smaller technological changes represent technological advances in our industry.	1	2	3	4	5	1	2	3	4	5	1	2	3	4	5

---

4 Technology stands for technological knowledge, procedures, materials and equipment. Compare the industry of your company relatively to other industries.

B.3. Evaluate performance of your company compared to your main competitors according to:
To answer use the following scale: 1 - Considerably worse than the main competitors, 2 - Worse than the main competitors, 3 - Same as main competitors, 4 - Better than the main competitors, 5 - Considerably better than the main competitors

	Whole company or A					B					C				
Number of modified, improved and completely new products in period 2004-2006.	1	2	3	4	5	1	2	3	4	5	1	2	3	4	5
We make quality products (from the viewpoint of use).	1	2	3	4	5	1	2	3	4	5	1	2	3	4	5
Research and development in the firm is advanced.	1	2	3	4	5	1	2	3	4	5	1	2	3	4	5
Number of available technological capabilities[5] inside the firm or through strategic partnerships.	1	2	3	4	5	1	2	3	4	5	1	2	3	4	5
We are good at predicting technological trends.	1	2	3	4	5	1	2	3	4	5	1	2	3	4	5
Obtaining information about changes of customer preferences and needs.	1	2	3	4	5	1	2	3	4	5	1	2	3	4	5
Acquiring real time information about competitors.	1	2	3	4	5	1	2	3	4	5	1	2	3	4	5
Establishing and managing long-term customer relations.	1	2	3	4	5	1	2	3	4	5	1	2	3	4	5
Establishing and managing long-term relations with suppliers.	1	2	3	4	5	1	2	3	4	5	1	2	3	4	5

---

5 Spectrum of technological knowledge that includes both practical and theoretical know-how, methods, processes, experience, technological devices and equipment. They represent firm's capacity to generate new knowledge, which is based primarily on experience and professional skills what sets technological capabilities apart from science.

	Whole company or A					B					C				
Good transfer of technological and marketing[6] knowledge among business units.	1	2	3	4	5	1	2	3	4	5	—	2	3	4	5
The intensity, quality and extent of research and development knowledge transfer in co-operation with strategic partners.	1	2	3	4	5	1	2	3	4	5	1	2	3	4	5
Product development is cost efficient.	1	2	3	4	5	1	2	3	4	5	1	2	3	4	5
Activities of business units are clearly defined in the corporate strategy of our firm.	1	2	3	4	5	1	2	3	4	5	1	2	3	4	5
Time needed to develop an improved product.[7]	1	2	3	4	5	1	2	3	4	5	1	2	3	4	5
Time needed to develop a new generation product.	1	2	3	4	5	1	2	3	4	5	1	2	3	4	5
Number of patents owned by the company.	1	2	3	4	5	1	2	3	4	5	1	2	3	4	5
Our firm substantially contributes to world trends in the industry.	1	2	3	4	5	1	2	3	4	5	1	2	3	4	5
Added value[8] of new products.	1	2	3	4	5	1	2	3	4	5	1	2	3	4	5
The company is cost efficient.	1	2	3	4	5	1	2	3	4	5	1	2	3	4	5
Quality-price relationship of our products is favourable for the buyer.	1	2	3	4	5	1	2	3	4	5	1	2	3	4	5

---

6 Marketing knowledge includes knowledge regarding recognizing, understanding and predicting needs and wishes of buyers as well as knowledge about competitors.
7 New modified product developed from existing technological knowledge, materials and processes.
8 Added value of new products equals the difference between sales and cost of goods sold (of new products).

C. R&D FUNCTION CHARACTERISTICS

➔ For questions from **C.1. to C.2.** choose **one answer** in each column.

C.1. Is R&D carried out internally or externally?[9]

	Whole or A	B	C
Only internal R&D[10]	☐	☐	☐
Mostly internal R&D, external to smaller extent	☐	☐	☐
Balanced internal and external R&D	☐	☐	☐
Mostly external R&D, internal to smaller extent	☐	☐	☐
Only external R&D	☐	☐	☐

C.1.1. If you have chosen one of the middle options, reply to the following question: **Which innovation has on average greater added value?**

	Whole or A	B	C
Innovation based on internal R&D	☐	☐	☐
Innovation based on external R&D	☐	☐	☐
Similar added value	☐	☐	☐

---

9 External R&D stands for contributions of strategic partner, e.g. suppliers, buyers, research institutions.

10 As internal R&D can be regarded also R&D carried out in a separate firm that is still a member of the same group to which belongs your company – it is how the group is organized. In this case mark if the company responsible for R&D is based in Slovenia? ☐ Yes ☐ No

C.2. What strategy does the company pursue product development?
Compare according to the number of innovation of a specific type.

> ⓘ
> **Strategy of imitation:** developed products share significant similarities with competition.
> **Strategy of innovation:** developed products are original and distinctively different from competition.

	Whole or A	B	C
Only strategy of imitation	☐	☐	☐
Mostly strategy of imitation, innovation to smaller extent	☐	☐	☐
Balanced strategy of imitation and innovation	☐	☐	☐
Mostly strategy of innovation, imitation to smaller extent	☐	☐	☐
Only strategy of innovation	☐	☐	☐

C.3. How would you describe the type of production employed at your company?
Please, provide the share of individual production type according to quantities produced during period 2004-2006.

	Whole or A	B	C
Customized production (made to order)	%	%	%
Production of series specified by the buyer	%	%	%
Production of standardized series (standard for the company)	%	%	%
Mass customization (modular production)	%	%	%

C.4. Innovation during period 2004-2006.

> ⓘ Innovation in a company consists of incremental and radical innovation. Innovation refers to product functions, features and shape.
> **Incremental innovation:** minor changes and improvements of products which are based on existing knowledge, technologies and materials.
> **Radical innovation:** based on original new knowledge and technologies. It is not merely an improvement of existing products but a new generation of considerably different products that is new to the firm and to the market. Often characterized by new different ways of use.

	Whole or A	B	C
Share of incremental innovation	%	%	%
Share of radical innovation	%	%	%

C.5. Newly introduced products during period 2004-2006.

	Whole or A	B	C
Share of sales attributed to improved products	%	%	%
Share of sales attributed to new generations of products	%	%	%

C.6. Mark importance of the following **contributions of incremental innovation** for your products? Innovations in the production processes are excluded.

→ On scale from 1 to 5 borderline value **1** means **Not important** and **5 Very important**.

	Whole or A	B	C
Improved product use	1 2 3 4 5	1 2 3 4 5	1 2 3 4 5
Improved product functionality	1 2 3 4 5	1 2 3 4 5	1 2 3 4 5
Lower production costs for your company	1 2 3 4 5	1 2 3 4 5	1 2 3 4 5
Improved appearance	1 2 3 4 5	1 2 3 4 5	1 2 3 4 5
Better company image[11]	1 2 3 4 5	1 2 3 4 5	1 2 3 4 5

C.7. **Number of awards won for innovation, quality and design of products between 2004 and 2006.** Including awards by institutions, associations, business partners, etc.

	Whole or A	B	C
Number of domestic awards			
Number of foreign awards			

---

[11] Incremental innovation presents for a company means of demonstrating its innovativeness and developing its image.

C.8. Did the company systematically enter new national markets (new countries) between years 2004 and 2006?

	Whole or A	B	C
Yes	☐ How many: ___	☐ How many: ___	☐ How many: ___
No	☐	☐	☐
Company does not systematically enter new markets but targets individual buyers regardless of their geographic location.	☐	☐	☐

C.9. Number of currently valid patents and models owned by the company?

	Whole or A	B	C
Number of patents			
Number of models protecting the appearance of products			

C.10. How many patents and models have you obtained in the past 3 years (2004-2006)?

	Whole or A	B	C
Number of patents for product innovation			
Number of patents for process innovation			
Number of models for visual appearance			

D. BASIC PERIODICAL COMPANY DATA

→ If your company belongs to a **group of interrelated companies** (by ownership) that work closely together along the supply chain in creating the value of the same final product, please provide **data for the group as a whole**.

D.1. Data will be provided for: ☐ **Your company**

☐ **Parent company in the group**
(if it is not your company)

☐ **Group** (consolidated data)

D.2. Data for period from 2002 to 2006:

	2002	2003	2004	2005	2006
Number of employees (annual average)					
Share of sales under own brand	%	%	%	%	%
R&D expenditure (absolute amount or as share of sales)					
Total costs of advertising and promotion (absolute amount or as share of sales)					

D.3 In what currency are stated absolute amounts?   ☐ 1000 SIT

☐ 1000 EUR

D.4. Provide data below only if given for the group.

	2002	2003	2004	2005	2006
Sales					
Costs of goods sold					
Export (absolute amount or as share of sales)					
EBIT					
Earnings					
Average gross monthly salary					
ROE	%	%	%	%	%
ROA	%	%	%	%	%

E. R&D EXPENDITURE AND FINANCING

E.1. R&D expenditure structure in period 2004-2006 as a share of total R&D expenditure.

	Share of the total
Basic research of new products and technologies	%
Research for improving existing products and technologies	%
Development of new generation products	%
Development of new production methods and processes	%
Laboratory activities	%
Total	**100 %**

E.2. Structure of financing sources of R&D in period 2004-2006 as a share of total R&D expenditure.

	Share of the total
Internal sources	%
Loans	%
Joint investment with domestic industrial partners	%
Joint investment with foreign industrial partners	%
Universities and research institutions	%
State funding	%
European Union funding	%
Other (explain): _____	%
Total	100 %

F. OWNERSHIP

F.1. Majority ownership: ☐ Domestic ownership

☐ Foreign ownership

F.2. Ownership structure for 2006.

	Share of ownership
State funds	%
Investment funds	%
Other companies	%
Banks	%
Minority owners	%
State of Republic of Slovenia and municipalities	%
Employee ownership	%
Management	%
Ex-employees, retired employees, relatives	%
Non-realized internal buyout	%
Other: _____	%
Total	100%

Respondent data

Current job position [ ]

Number of years in the company [ ]

Number of years at the current position [ ]

Number of years in the company
(including other companies) [ ]

THANK YOU VERY MUCH FOR YOUR COOPERATION.

Appendix D

Dendrogram

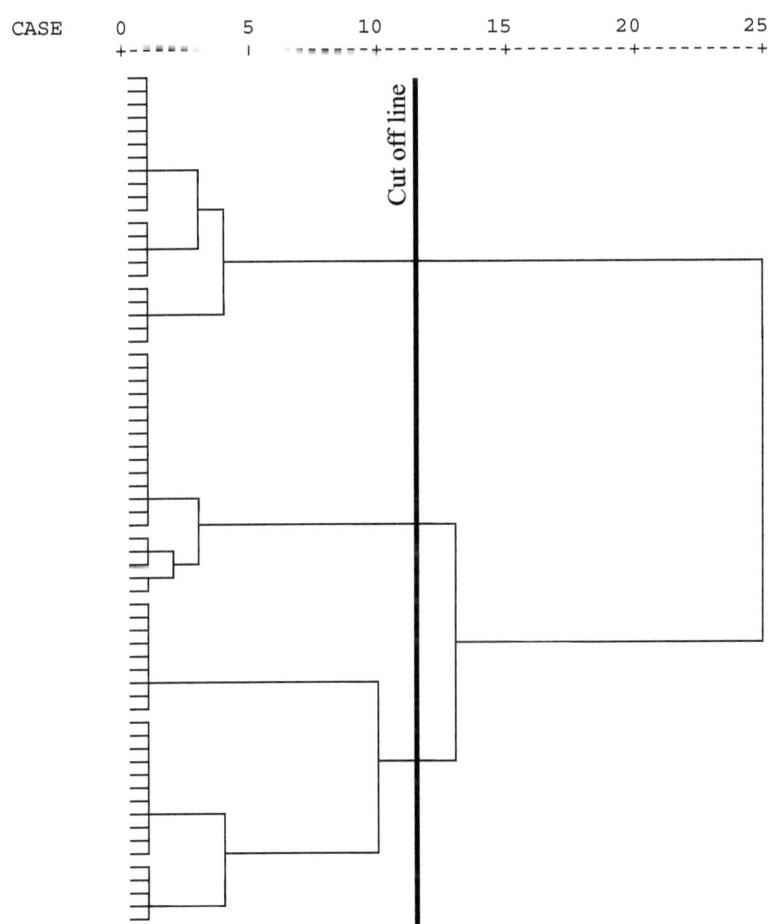

Note: Ward's procedure; Squared Euclidean Distance. Dendrogram obtained using SPSS program.

Appendix E

Comparison of object classification with hierarchical Ward's procedure and K-means method into 3 clusters

N	Ward	K-means	N	Ward	K-means	N	Ward	K-means
1	1	1	24	1	1	47	3	3
2	1	1	25	1	1	48	3	3
3	1	1	26	2	2	49	3	3
4	1	1	27	2	2	50	3	3
5	1	1	28	2	2	51	3	3
6	1	1	29	2	2	52	3	3
7	1	1	30	2	2	53	3	3
8	1	1	31	2	2	54	3	3
9	1	1	32	2	2	55	3	3
10	1	1	33	2	2	56	3	3
11	1	1	34	2	2	57	3	3
12	1	1	35	2	2	58	3	3
13	1	1	36	2	2	59	3	3
14	1	1	37	2	2	60	3	3
15	1	1	38	2	2	61	3	3
16	1	1	39	2	2	62	3	3
17	1	1	40	2	2	63	3	3
18	1	1	41	2	2	64	3	3
19	1	3	42	2	2	65	3	3
20	1	1	43	2	2			
21	1	1	44	2	2			
22	1	1	45	3	3			
23	1	1	46	3	3			

Note: Object number 19 was differently classified depending on the clustering method employed.

Appendix F

PLS structural model analysis for the incremental innovation model of innovative performance

Values of composite reliability for all constructs exceed 0.70, thereby confirming internal consistency reliability (Table F.1). In Table F.2 are listed cross loadings higher than the mean which is 0.5963. The minimum cross loading value of the proposed indicators in the model is 0.6479. All latent variables thus appear to be well correlated with their own indicators, thereby, speaking in favour of the convergent validity. The lowest value of AVE is 0.5586. The criterion of discriminant validity is also satisfied with square root of AVE of each latent variable shown on the diagonal in Table F.1 exceeding all of the correlation coefficients stated below. The value of R^2 is 0.571.

Table F.1: Composite reliability, correlation matrix and the square roots of AVE

	Composite reliability	TC	MC	CC	IP
TC	0.9197	*0.8902*			
MC	0.8450	0.6531	*0.7622*		
CC	0.8342	0.6528	0.7153	*0.7473*	
IP	0.8085	0.6756	0.6495	0.6718	*0.8237*

Note: The square roots of AVE are in the diagonal in italics.
Below the diagonal are correlation coefficients.

Table F.2: Cross loadings between manifest and latent variables

Indicators	TC	MC	CC	IP
RD_ADVAN	**0.8518** **(19.255)**			
TECH_CAP_NQ	**0.9109** **(42.992)**	0.6024		0.6761
TECH_TREND_F	**0.9070** **(32.517)**	0.6279	0.6071	
INFO_CUST	0.6140	**0.8372** **(23.221)**	0.6205	0.6004
INFO_COMP		**0.6485** **(6.815)**		
CUST_RELAT		**0.8717** **(29.054)**	0.6166	
SUPP_RELAT		**0.6654** **(8.458)**		
TECH_MRKT_KN			**0.7983** **(12.064)**	
RD_STP			**0.7522** **(18.530)**	
RD_COST_EFF	0.6547		**0.6479** **(6.094)**	
ACT_STRAT		0.6865	**0.7821** **(11.380)**	
TIME_IMPR				**0.8223** **(12.259)**
QUAL_PROD				**0.8251** **(16.799)**

Note: T-values are stated in parentheses for those indicators that belong to a designated latent variable in the model. All significant at P<0.001.

Appendix G

PLS structural model analysis for the radical innovation model of innovative performance

For the model of innovative performance referring to radical innovation, again internal consistency reliability, convergent validity and discriminant validity are confirmed (Table G.1 and Table G.2). The lowest value of composite reliability is 0.8330, which is considerably higher than the threshold value of 0.70 (Table G.1). In Table 18 are listed cross loadings with values above the mean value 0.5942. The lowest cross loading is 0.6409 and above the cut-off point at 0.60. The lowest value of AVE is 0.5581 and again all square root values of AVE exceed the correlation coefficient stated below them in Table G.1. The value of $R2$ is 0.577.

Table G.1: Composite reliability, correlation matrix and the square roots of AVE

	Composite reliability	TC	MC	CC	IP
TC	0.9200	*0.8905*			
MC	0.8449	0.6535	*0.7619*		
CC	0.8338	0.6618	0.7147	*0.7470*	
IP	0.7825	0.6852	0.6256	0.6899	*0.8017*

Note: The square roots of AVE are in the diagonal in italics.
Below the diagonal are correlation coefficients.

Table G.2: Cross loadings between manifest and latent variables

Indicators	TC	MC	CC	IP
RD_ADVAN	**0.8592** **(24.328)**			
TECH_CAP_NQ	**0.9034** **(39.869)**	0.6042		0.6487
TECH_TREND_F	**0.9083** **(32.894)**	0.6287	0.6125	0.6006
INFO_CUST	0.6151	**0.8396** **(23.246)**	0.6191	
INFO_COMP		**0.6646** **(8.066)**		
CUST_RELAT		**0.8665** **(22.610)**	0.6196	
SUPP_RELAT		**0.6520** **(7.552)**		
TECH_MRKT_KN		0.5957	**0.7888** **(10.258)**	
RD_STP			**0.7730** **(19.357)**	
RD_COST_EFF		0.6539	**0.6409** **(5.549)**	
ACT_STRAT			**0.7758**	
		0.6827	**(10.385)**	
TIME_NEW				**0.7953** **(10.909)**
QUAL_PROD				**0.8081** **(11.147)**

Note: T-values are stated in parentheses for those indicators that belong to a designated latent variable in the model. All significant at P<0.001.

Appendix H

PLS structural model analysis for the trend setting/market leadership model of innovative performance

Internal consistency reliability, convergent validity and discriminant validity were confirmed also for the third model referring to trend setting in the frame of innovative performance. The lowest composite reliability value is 0.7623. The lowest AVE value is 0.5584. From Table H.1 it can also be concluded that all square root values of AVE of the latent variables exceed the values of their correlations with other included latent variables. The value of R2 is 0.516. In Table H.2 are listed cross loadings above the mean value of 0.5825, the lowest being 0.6240.

Table H.1: Composite reliability, correlation matrix and the square roots of AVE

	Composite reliability	TC	MC	CC	IP
TC	0.9197	*0.8902*			
MC	0.8450	0.6536	*0.7622*		
CC	0.8336	0.6553	0.7285	*0.7472*	
IP	0.7623	0.6322	0.5937	0.6663	*0.7856*

Note: The square roots of AVE are in the diagonal in italics.
Below the diagonal are correlation coefficients.

Table H.2: Cross loadings between manifest and latent variables

Indicators	TC	MC	CC	IP
RD_ADVAN	**0.8470**			
	(21.466)			
TECH_CAP_NQ	**0.9104**			
	(42.796)	0.6016		0.6249
TECH_TREND_F	**0.9118**			
	(37.475)	0.6275	0.6173	
INFO_CUST		**0.8365**		
	0.6147	**(21.717)**	0.6252	
INFO_COMP		**0.6414**		
		(6.270)		
CUST_RELAT		**0.8733**		
		(29.616)	0.6311	
SUPP_RELAT		**0.6712**		
		(8.235)		
TECH_MRKT_KN			**0.7953**	
		0.5953	**(11.400)**	
RD_STP			**0.7439**	
	0.6539		**(12.408)**	
RD_COST_EFF			**0.6240**	
			(5.187)	
ACT_STRAT			**0.8113**	
		0.688	**(14.610)**	0.5855
TRENDS				**0.7241**
				(6.814)
QUAL_PROD				**0.8427**
			0.5950	**(12.765)**

Note: T-values are stated in parentheses for those indicators that belong to a designated latent variable in the model. All significant at P<0.001.